Neonatal Equipment

Everything that you would like to know!

Fifth Edition

Neonatal Equipment

Everything that you would like to know!

Fifth Edition

Chief Editor
Ashok K Deorari
MD, Diplomate NB, FAMS, Fellow NNF

Professor and Head, Department of Pediatrics
WHO Collaborating Centre for Newborn Training and Research
All India Institute of Medical Sciences, New Delhi

Co-Editors
Vinod K Paul
MD, PhD, FAMS, FNASc, FASc, FNA

Professor, Department of Pediatrics
All India Institute of Medical Sciences, New Delhi
Member, NITI Aayog
National Institution for Transforming India
Government of India, New Delhi

Anu Sachdeva
MD, DM

Assistant Professor
Department of Pediatrics
WHO Collaborating Centre for Newborn Training and Research
All India Institute of Medical Sciences, New Delhi

CBS Publishers & Distributors Pvt Ltd

New Delhi • Bengaluru • Chennai • Kochi • Kolkata • Mumbai
Bhopal • Bhubaneswar • Hyderabad • Jharkhand • Nagpur • Patna • Pune • Uttarakhand • Dhaka (Bangladesh)

Disclaimer

Science and technology are constantly changing fields. New research and experience broaden the scope of information and knowledge. The editors have tried their best in giving information available to them while preparing the material for this book. Although all efforts have been made to ensure optimum accuracy of the material, yet it is quite possible some errors might have been left uncorrected. The publisher, the printer and the editors will not be held responsible for any inadvertent errors, omissions or inaccuracies.

ISBN: 978-93-85915-78-9

Copyright © Editors and Publisher

Fifth Edition 2017
 Reprint 2019
First Edition 1999
Second Edition 2001
Third Edition 2006
Fourth Edition 2010

All rights reserved. No part of this book may be reproduced or transmitted in any form or by any means, electronic or mechanical, including photocopying, recording, or any information storage and retrieval system without permission, in writing, from the Editors and the publisher.

Published by Satish Kumar Jain and produced by Varun Jain for
CBS Publishers & Distributors Pvt Ltd
4819/XI Prahlad Street, 24 Ansari Road, Daryaganj, New Delhi 110 002, India.
Ph: 23289259, 23266861, 23266867   Fax: 011-23243014   Website: www.cbspd.com
                                                              e-mail: delhi@cbspd.com; cbspubs@airtelmail.in

Corporate Office: 204 FIE, Industrial Area, Patparganj, Delhi 110 092
Ph: 4934 4934   Fax: 4934 4935   e-mail: publishing@cbspd.com; publicity@cbspd.com

Branches

- **Bengaluru:** Seema House 2975, 17th Cross, K.R. Road,
  Banasankari 2nd Stage, Bengaluru 560 070, Karnataka
  Ph: +91-80-26771678/79   Fax: +91-80-26771680   e-mail: bangalore@cbspd.com
- **Chennai:** 7, Subbaraya Street, Shenoy Nagar, Chennai 600 030, Tamil Nadu
  Ph: +91-44-26680620, 26681266   Fax: +91-44-42032115   e-mail: chennai@cbspd.com
- **Kochi:** 42/1325, 1326, Power House Road, Opposite KSEB Power House,
  Ernakulam 682 018, Kochi, Kerala
  Ph: +91-484-4059061-65   Fax: +91-484-4059065   e-mail: kochi@cbspd.com
- **Kolkata:** 6/B, Ground Floor, Rameswar Shaw Road, Kolkata 700 014, West Bengal
  Ph: +91-33-22891126, 22891127, 22891128   e-mail: kolkata@cbspd.com
- **Mumbai:** 83-C, Dr E Moses Road, Worli, Mumbai 400018, Maharashtra
  Ph: +91-22-24902340/41   Fax: +91-22-24902342   e-mail: mumbai@cbspd.com

Representatives

- Bhopal   0-8319310552   • Bhubaneswar   0-9911037372   • Hyderabad   0-9885175004   • Jharkhand   0-9811541605
- Nagpur   0-9421945513   • Patna   0-9334159340   • Pune   0-9623451994   • Uttarakhand 0-9716462459
- Dhaka (Bangladesh) 01912-003485

Printed at Goyal Offset Printers, GT Karnal Road, Industrial Area, Delhi, India

to

National Neonatology Forum
for spearheading the newborn care movement in India

Foreword

The passive nurse-dominated approach for the care of newborn babies has now been replaced by more active and aggressive technology-driven monitoring and management strategies for the care of preterm and sick newborn babies. To meet the increasing technology needs of the neonatal intensive care units, there has been a revolution in our country for indigenous fabrication of a large number of both simple and complex neonatal electronic equipment. Despite their ready availability, most consumers are frustrated by their exorbitant prices, substandard quality, frequent breakdowns, unsatisfactory preventive maintenance and delay in their repair due to lack of spares or unsatisfactory infrastructure of vendors. It is an unfortunate reality that at a given point of time, a large number of units of life-support intensive care equipment are malfunctioning or are lying out of order in most neonatal units in the country.

There was a long-standing felt need to have a comprehensive, updated and user-friendly manual of neonatal equipment for the benefit of the harassed consumers. The *Neonatal Equipment: Everything that you would like to know!* is indeed a milestone publication in neonatology. I am confident it would be of great practical benefit to the pediatricians and planners to make a rational choice of neonatal equipment from reliable and credible local vendors and manufacturers. I do hope that the manual would meet the lofty objectives and aspirations of the editors so that every neonatal unit in our country would be equipped with appropriate, reliable, sturdy and functional equipment in the next millennium.

New Delhi
14th November, 1999

Meharban Singh
MD, FAMS, FIAP, FAAP
Former Professor and Head
Department of Pediatrics
All India Institute of Medical Sciences
New Delhi

Contributors

Amanpreet Sethi
DM Fellow
Department of Pediatrics
All India Institute of Medical Sciences
New Delhi 110029

Anu Sachdeva
Assistant Professor
Department of Pediatrics
All India Institute of Medical Sciences
New Delhi 110029

Arvind Shenoi
Consultant Neonatologist
Cloud Nine
Manipal Hospital
Bangalore 560017, Karnataka

Ashok K Deorari
Professor, Neonatal Division
Department of Pediatrics
All India Institute of Medical Sciences
New Delhi 110029

Deepak Chawla
Associate Professor
Department of Pediatrics
Government Medical College, Sector 32
Chandigarh 160030

G Girish
Consultant Neonatologist
Apollo Hospital
Mysore, Karnataka

Gouri Rao Passi
Consultant Pediatrician
Choitram Hospital
Indore 452001, Madhya Pradesh

Indrani Sarkar
DM Fellow
Department of Pediatrics
All India Institute of Medical Sciences
New Delhi 110029

Jagjit Singh
DM Fellow
Department of Pediatrics
All India Institute of Medical Sciences
New Delhi 110029

Jeeva Sankar
Assistant Professor
Neonatal Division
Department of Pediatrics
All India Institute of Medical Sciences
New Delhi 110029

JP Dadhich
Consultant Pediatrician
Sunderlal Jain Hospital
Ashok Vihar, Delhi 110052

Meena Joshi
Nurse Educator
Department of Pediatrics
All India Institute of Medical Sciences
New Delhi 110029

Naveen Sankhyan
Assistant Professor
Postgraduate Institute of Medical Education
and Research
Chandigarh 160 012

PK Rajiv
Incharge, Neonatologist
Amrita Institute of Medical Sciences
Kochi 17, Kerala

Praveen Kumar
Professor
Neonatal Division
Department of Pediatrics, PGIMER
Chandigarh 160012

Prem N Sheth
Consultant Neonatologist
Department of Pediatrics
Bombay Hospital
Mumbai 400020, Maharashtra

Rahul Verma
Consultant Pediatrician
Neonatologist Bombay Hospital
Mumbai 400020, Maharashtra

Rajiv Aggarwal
Consultant Neonatologist and Pediatrician Intensivist
and HOD of Narayana Children's Hospital
Bangalore 560 099

Rakesh Lodha
Additional Professor
Department of Pediatrics
All India Institute of Medical Sciences
New Delhi 110029

Ramesh Agarwal
Additional Professor
Department of Pediatrics
All India Institute of Medical Sciences
New Delhi 110029

Shankar Narayan
Pediatrician, Indian Navy
Colaba, Mumbai

Sindhu S
DM Fellow
Department of Pediatrics
All India Institute of Medical Sciences
New Delhi 110029

Sourabh Dutta
Professor
Department of Pediatrics, PGIMER
Chandigarh 160012

Sushma Nangia
Professor
Department of Pediatrics
Kalawati Saran Childrens Hospital
New Delhi 110001

Suvasini Sharma
Assistant Professor
Department of Pediatrics
Lady Hardinge Medical College
New Delhi 110001

Swarna Rekha
Ex-Professor and Head
Department of Pediatrics
St. John's Medical College
Sarjapur Road, Bangalore 560034

Tanushree Sahoo
DM Fellow
Department of Pediatrics
All India Institute of Medical Sciences
New Delhi 110029

Vinod K Paul
Professor and Head
Department of Pediatrics
All India Institute of Medical Sciences
New Delhi 110029

Preface to the Fifth Edition

Modern day neonatal intensive care necessitates the use of equipment for optimum care of sick newborn babies. Equipment alone, however, does not necessarily improve the care of the high-risk neonates. Availability of space, maintenance of asepsis and presence of adequate nursing staff are important prerequisites before sophisticated equipment is acquired. Maintenance of the existing equipment is more important than acquisition of new gadgets. We feel that as a pediatrician, it is essential that we are aware about what to buy, where to buy, and how to use the common neonatal equipment. With this aim we are privileged to bring before you the fifth edition of most sought after user-friendly book on key neonatal equipment.

This book is a compilation contributed by a panel of distinguished neonatologists of India. Each chapter deals in detail with the principles of working of equipment and indications and practical considerations. A comprehensive table in each chapter provides information on the available models along with approximate cost and names of the dealers in the Indian market. To clarify doubts in the mind of the users, frequently asked questions and answers are discussed at the end of each chapter. This revised edition has been brought out on the advice of several pediatricians, but Dr Shankar Narayan, trainee fellow in neonatology from the Indian Navy at AIIMS, deserves special mention. A number of suggestions made by him in the earlier edition have been incorporated. Five new equipments, viz. inhaled NO, heated humidified high flow nasal cannula, otoacoustic emission, cooling machine, T-piece resuscitator, have been included. Chapter on how to select appropriate equipment and write specifications for tender purposes have been rewritten to address the needs of user. The information on dealers and their addresses have been updated. We plan to disseminate this book throughout the country among those who care for the newborn babies. This we hope would go a long way to ensure better utilization of equipment and enhance newborn survival.

We are grateful to the equipment manufacturers and suppliers for providing details about their products. The editors are indebted to the contributors for their outstanding write-ups and to Dr Jagjit Singh, Dr S Sindhu, Dr Amanpreet Sethi, Dr Mukul Kumar Mangla, Dr Tanushree Sahoo, Dr Pratima Anand, Dr Srikanth Kulkarni, DM trainee fellows, for reading the galley proofs. We would like to express appreciation to Mrs Tessy Varghese for diligently typing the manuscript and Mr YN Arjuna, Senior Vice President—Publishing Editorial and Publicity, CBS Publishers & Distributors Pvt Ltd, for publishing the book under his expert supervision. Kindly note, that your comments and suggestions are invaluable. We would appreciate your comments, suggestions and inputs for future editions of the book.

New Delhi
15th August, 2016

Ashok K Deorari
Vinod K Paul
Anu Sachdeva

Contents

Foreword by Meharban Singh — *vii*
Contributors — *ix*
Preface to the Fifth Edition — *xi*

A. INTRODUCTION

1. Procurement, Use and Maintenance of Neonatal Equipment: Practical Considerations — 3

B. EQUIPMENT FOR THERMAL CONTROL

2. Radiant Warmers — 11
3. Incubators — 17
4. Transport Incubators — 25

C. EQUIPMENT FOR MONITORING

5. Pulse Oximeter — 33
6. Apnea Monitors — 43
7. Blood Pressure Monitors — 48
8. Transcutaneous Bilirubinometer — 55
9. Thermometers — 59
10. Weighing Scales — 64
11. Transcutaneous Blood Gas Monitors — 68
12. Capnograph (End-Tidal Carbon Dioxide Monitors) — 75
13. Ultrasound Machine — 82
14. Cerebral Function Monitor — 87
15. BERA Phone: Automated Auditory Brainstem Evoked Response Audiometry — 94
16. Otoacoustic Emission (OAE) Machine — 99

D. EQUIPMENT FOR THERAPY/TREATMENT PURPOSES

17. Phototherapy Units — 107
18. Infusion Pumps — 117
19. Cooling Equipment for Therapeutic Hypothermia — 122

E. EQUIPMENT FOR MONITORING THERAPY

20.	Oxygen Analyzer	133
21.	Fluxmeter (Irradiance Meter for Phototherapy Unit)	136
22.	Transilluminator	139

F. EQUIPMENT FOR BIOCHEMICAL AND LABORATORY MEASUREMENT

23.	Glucose Estimation Instruments	145
24.	Clinical Refractometer	150
25.	Spectrometric Bilirubin Analyzer (Micro-method for Serum Bilirubin Estimation)	154
26.	Laboratory Microcentrifuge	159
27.	Laminar Airflow System	164

G. LIFE-SAVING EQUIPMENT

28.	Self-inflating Bag: Manual Resuscitator	173
29.	Oxygen Concentrator	179
30.	Continuous Positive Airway Pressure Machine	185
31.	Heated Humidified High Flow Nasal Cannula Therapy	195
32.	Neonatal Ventilators	198
33.	Inhaled Nitric Oxide Delivery Systems	205

H. MISCELLANEOUS

34.	Resuscitation Trolley	213
35.	T-piece Resuscitator	216
36.	Suction Machine	220
37.	Breast Pump	224

I. APPENDIX

38.	Steps for Equipment Selection and Writing Specification	231
39.	List of Indian Equipment Dealers	240
40.	List of Imported Equipment Dealers	245
	Index	249

Section

A. Introduction

1. Procurement, Use and Maintenance of Neonatal Equipment: Practical Considerations

Chapter 1

Procurement, Use and Maintenance of Neonatal Equipment: Practical Considerations

Optimum care of sick and small neonates necessitates the use of biomedical equipment. These gadgets range from simple thermometers to microprocessor-based incubators and ventilators. Some equipment is manufactured in the country, while other needs to be imported. However, all of them entail quite a cost for a poor country like ours. Pediatricians endeavoring to establish neonatal units, often with limited resources and in the face of competing priorities, find it difficult to decide the right category and make of equipment. Once acquired, the equipment may remain idle or in disrepair for ages due to trivial reasons or lack of effort on the part of the supplier or the user or both. Indeed, equipment maintenance is a nightmare for any physician.

We would now be able to take up some important questions regarding neonatal equipment.

What equipment do I need to acquire?

This is the most important question. The equipment you choose should be priority-based, rational and affordable.

The scope and stage of the neonatal care services one proposes to develop determines the package of equipment you require. For instance, resuscitation equipment (suction machine, resuscitation bag, oxygen and laryngoscope), thermometer, oxygen hoods and a weighing scale are usually the first bunch of equipment required to start a newborn service. At this stage, one will also require plenty of disposable items such as feeding tubes, syringes, katori–spoons, intravenous cannulae and small volume infusion sets. One may then supplement them with radiant warming system(s) or incubators. The next logical equipment to buy are perhaps a phototherapy unit and a pulse-oximeter. The latter is an excellent monitoring device that provides not only oxygen saturation in the blood but also the heart rate. It gives a warning in the event of hypoxia, bradycardia and tachycardia.

The issue of rationality in deciding the equipment needs elaboration. It is irrational to think of buying a ventilator before ensuring provision of pulse oximetry and arterial blood gas (ABG) estimation. In addition, there must be sufficient staff round the clock to manage the ventilated neonates. Likewise, it is irrational to have pulse oximeter but no oxygen hood to deliver oxygen.

Which indigenous equipment is of acceptable quality?

In recent years, good quality indigenous equipment of certain categories has become available. These include resuscitation bags, weighing scales (up to 1–10 gm sensitivity), oxygen head boxes and radiant warming (open care) systems. On the other hand, locally made phototherapy units, incorporating white tubelights do not provide adequate irradiance but recently blue light sleek light emitting diode or compact fluorescent lights which are lightweight portable are available in the market. Those using halogen bulbs, although effective, are bulky, tend to overheat the baby, and their bulbs have a short life. Incubators and CPAP machines made in the country are also not of satisfactory quality as yet. Currently even infusion pumps and pulse oximeters are being assembled in the country but they need to improve the quality of the products.

Is it better to buy indigenous or imported equipment?

For obvious reasons, this choice is unavailable for equipment like non-invasive blood pressure monitors, ventilators, blood gas analyzers and bilirubin analyzers which are not manufactured in India. For that equipment which is manufactured in India and is of reasonable quality, it is better to buy an Indian product. The cost is usually much lower than the imported versions, although, often they do not have impressive looks and are not as sturdy. For the cost of one imported open care system, you can acquire 3 to 4 indigenous ones which is a good deal in practical terms.

Do you think there should be accreditation for medical equipment in India?

Yes, accreditation and standardization are extremely necessary to monitor the quality of the mushrooming of small time equipment manufacturers in India. Today, anybody can start manufacturing equipment, like any other small business, without any permission or license. By the same norm, the quality of equipment sold by the MNCs needs to be filtered. Also India should not become a dumping ground for other countries. In the US, Japan and China, one needs to have FDA, PMDA and SDA approvals respectively, to sell medical equipment. Similarly, no medical equipment can be installed in the European Union countries, unless it carries the CE mark.

Are the domestic manufacturers losing out to the MNCs? Is there a clash of clientele?

No, there is no clash. The MNCs target the corporate hospitals and super-specialty hospitals, mainly institutes, where price is less of a concern for buying equipment. Only 0.5 percent of the population, i.e. the high income group can afford the treatment offered by corporate and super-specialty hospitals. The indigenous industry has customers mainly in nursing homes, clinics as well as private and government hospitals.

What is the scope of the growth of the medical equipment market?

As of today, the 5000 crore diagnostic and imaging equipment industry, to a large extent dominated by MNCs, like GE, Philips Medical Systems, Siemens, etc. is growing at a rapid pace of 15–20% annually. In the health care, medical equipment market segment is one of the top growth markets in India. It is estimated to be $922 million in 2010 and is forecast to reach $1,693 million by 2014. The market is highly dependent on imports, with more than 50% of the demand fulfilled by imports.

What protocol recipient needs to follow while accepting equipment from donor?

Following checklist will help

1. Prepare an equipment checklist and check whether the equipment to be received conforms to the set checklist. This checklist should be made considering factors such as name/brand of the equipment, country of origin, company track record, installations if any around, import legalities involved, *octroi*, sales tax if any applicable, technical specifications, such as humidity, temperature, voltage, frequency, current required, size, weight, etc.

2. Whether equipment is provided with the required accessories and reasonable quantity of spares and consumables

essential for the basic operation of the equipment. This should take into account the "lead period" (i.e. period between placing an order and receipt of spare parts).
3. Whether equipment can be fully supported with spares/accessories in the recipient's country/area.
4. Whether equipment is provided with a complete manual of operations and technical manual, for its use by the end user and the technical team for maintenance.
5. Ensure availability of the required trained staff (clinical and technical) for the installation, operation and maintenance of the equipment.
6. Keep maintenance costs, import costs, taxes if any, in mind, etc.
7. All new equipment must be accompanied by documents of warranty/guarantee.
8. Availability of required space for the equipment, keeping in mind factors such as accessibility, environmental conditions (temperature, humidity, etc.), utilities such as required power supply, gas supply, type of water if required, etc.
9. Availability of the support services required for the optimal functioning of the equipment.
10. Involve technical departments and the end user for the type of equipment required, its use, application, etc.

Enumerate Obligations of Genuine Donor for Humanitarian Cause

Following stated guidelines if implemented and ensured are sure to assist an effective donation of the medical equipment thereby complying with its basic reason of donation: "Optimal use of functional ability of the equipment".

1. Ensure that the equipment supplied is clinically, technologically, ergonomically and economically appropriate for the recipient.
2. Request for the equipment checklist set by the recipient and verify whether the equipment to be supplied conforms to the same. Both the donor and the recipient should reach an agreement on all conditions before the shipping of the equipment.
3. A basic list of all components must be provided and the "life expectancy" of the equipment should be clearly stated.
4. Ensure that the equipment supplied is fully equipped with the required accessories, spares, manuals and other related documents.
5. Supply an initial requirement of essential spares and consumables as agreed upon.
6. Provide the manufacturer's contact details to ensure easy accessibility by the recipient in case of any queries, requirements (of spares/consumables), etc.
7. Provide the contact details of the authorized distributor/local dealer if any.
8. Ensure proper packaging and provide the complete packaging list to the recipient.
9. Packaging should be strong and sturdy to avoid damage during transportation.
10. Ensure immediate delivery of all the shipping documents to the recipient to avoid delay in receiving the equipment, due to unavailability of essential shipping documents required at the recipient end.
11. Understand the import regulations/liabilities like taxes, import duties, etc. and the regulations for donated equipment if any of the recipients' country. Ensure the ability of the recipient to conform to such liabilities/regulations.

How should one select a particular brand of equipment?

This choice should depend on a careful scrutiny of all the available brands.

1. Make sure that the given equipment is technically capable of providing the required care, has reasonable precision of controls, provides the required alarms, is

sturdy and sleek, runs on 220 volts, has good wheels, and so on. Go for equipment holding certification of standards (ISI, ISO, FDA, PMDA, etc.).
2. Check about the functioning of the equipment from friends and colleagues who have used the same equipment. Also confirm the track record of the company from the users. A company having a headquarter or an agent near one's town may be in a better position to provide satisfactory service.
3. Bargain for the price because affordability needs to be taken into consideration in selecting one brand versus the other.
4. For indigenous products especially the life-saving machines be very sure to buy safe and reliable machines. Unfortunately in India we do not have strict laws which can have check on marketing of these products.

While buying equipment, what other practical points should be kept in mind?

1. Ask for sufficient warranty period (say, 2 years with spares, plus 3 years without spares) in writing, specifying the frequency of preventive maintenance checks during that period.
2. Ask for reusable and not disposable parts/accessories for items such as incubator probes or BP cuffs. Include additional accessories to last 2 to 5 years. For instance, extra tubelights for phototherapy units, extra temperature probes for the incubator, extra cuffs for BP monitors or probes of pulse oximeters should be procured. These should be included in the initial order.
3. Take commitment from the supplier to install the equipment by a specific date and to train one's staff in the use of the equipment through a specified number of training sessions. Also, insist on providing users' manuals.
4. Take commitment from the supplier to provide service for at least 7 years (which is the usual life of electromedical equipment) in writing, specifying the "uptime" (time for which the equipment shall be functional, usually taken as 90%) and "response time" (time taken to examine the equipment upon receiving a complaint, usually less than 24 hours).
5. For imported equipment, make sure that the local supplier possesses an authorization from the parent company. For highly expensive equipment, a counter-guarantee of service should also be taken from the foreign principals.
6. Check with the supplier if any preinstallation facilities (such as wall-mounts) are required.
7. It is advisable to run all equipment on stabilized voltage to guard against damage due to fluctuation in voltage which is so common in our country.

How does one maximize the use of available equipment?

Here are some suggestions for getting the best out of your equipment:
1. Train all concerned staff in proper handling and use of the given equipment. Reinforce this by simple placards explaining the steps of correct use as well as how to switch it off and keep it properly when not in use.
2. Provide a copy of users' manual in the unit and display simple trouble-shooting tips.
3. Include the demonstration of use and trouble-shooting of the equipment as a part of the postgraduate assessment and examination. This will help the residents to master their correct use and care.
4. Insist on regular disinfection after use of the equipment on a baby.
5. Encourage at least one person in one's team to master the care of the equipment. He/she can help solve the day to day problems in their use.

How does one ensure good maintenance and service of equipment?

One has to be obsessed with ensuring full functioning of the equipment; only then can one ensure that most of the equipment under one's charge remains in working order. Hardly any suppliers, whether of the indigenous or the imported equipment, will ever come up to expectations in providing satisfactory service. Be prepared to be after them to get one's work done.

Insist on preventive maintenance visits. Formulate annual or biannual maintenance contracts and earmark funds for spares and accessories. It is advisable to maintain an equipment register in which the events pertaining to the equipment (from acquisition, installation, cost, to breakdowns) is recorded. Keep record of when the complaint was lodged. Do send a written complaint if verbal/telephonic complaint is not responded to in a reasonable time. Do not hesitate to write to the parent company if there is unacceptable delay in repairs. Documentation is a deterrent to the companies and can be used in due course to inform the principals or seek redressal at the consumer courts. Tell one's friends about the companies with callous attitude so that they do not become their victims. Raise these issues at annual meetings of the professional organizations such as National Neonatology Forum. One can write to equipment committee of NNF. For details visit website www.nnfequipment.org Do not give up till your equipment is repaired and made fully functional! Do not we do that for our personal vehicle!

Section B. Equipment for Thermal Control

2. Radiant Warmers
3. Incubators
4. Transport Incubators

Chapter 2

Radiant Warmers

Lack of attention to thermoregulation continues to be a cause of unnecessary deaths in the neonatal population. Maintaining a stable body temperature is essential to ensure optimal growth. If temperature is maintained, caloric expenditure and oxygen consumption is minimal. Newborn babies, in particular the preterm and the low birth weight are exquisitely predisposed to hypothermia. No other equipment is identified more with the special care of newborn babies than the radiant warmers.

They provide intense source of radiant heat energy. They also reduce the conductive losses by providing a warm microenvironment surrounding the baby. The radiant warmer (*also called open care system*) was developed as an 'open incubator' that ensures ready access to the baby. The overhead quartz/ceramic heating element produces heat which is reflected by the parabolic reflector onto the baby on the bassinet. The quantity of heat produced is displayed in the heater output display panel. Temperature selection knobs select the desired skin temperature. This information is processed by the microprocessor inside the control panel and matched against the actual temperature of the baby. If the temperature of the baby is lower than the set temperature, the microprocessor will send feedback to the rod heater to increase the heat output till the baby's temperature reaches the set temperature. At this point, the heater output will be reduced. This system in which the heater output is determined automatically based on skin temperature information is called *servo system*. Servo system is the preferred method of running the open care system. The heat output from the heating rod could also be increased or decreased manually. This is done by the heater output control knobs. This is called the *manual mode* of operation. Whenever the baby's temperature rises by more than 0.5°C above the set temperature, a visual/audible alarm is activated in the servo mode. Caregiver can pay attention to sort out the fault. Often this occurs when the skin probe comes off the baby.

Parts of Open Care System

- *Bassinet*: For placing the neonate.
- *Quartz/ceramic rod*: Provides radiant heat.
- *Skin probe*: When attached to the baby's skin and displays skin temperature.
- *Control panel*: It has a collection of display and control features/knobs.
- *Heater output display*: Indicates how much is the heater output.
- *Heater output control knobs*: For increasing or decreasing the heater output manually.
- *Temperature selection panel*: Select either set temperature or skin temperature.
- *Temperature selection knobs*: Select a desired set temperature.
- *Temperature display*: Display temperature as selected, either of the baby's skin (via skin probe) or the set temperature.
- *Mode selector*: Selects manual or servo mode.

Neonatal Equipment

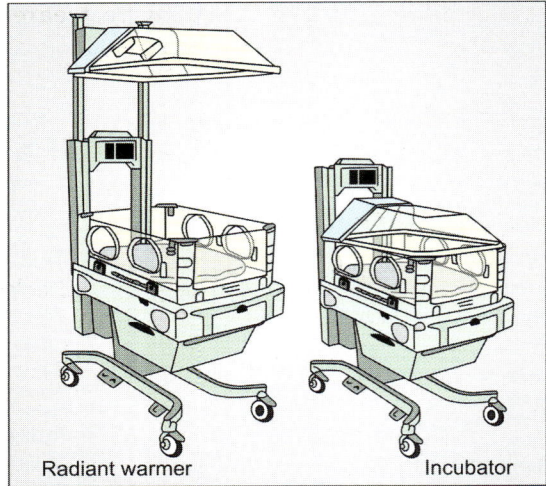

Radiant warmer   Incubator

Fig. 2.1: A typical radiant warmer and an incubator

The heating element (silicon quartz/infrared/ceramic/quartz crystal), the control panels (electronic/electrical/microprocessor based) and alarms (air over temperature/skin over temperature/air sensor fail/power failure, etc.) forms the basic unit of all the warming devices. Power consumption is low around 750 watts. In good equipment, temperature stability is usually with an accuracy of ± 0.5°C.

Steps for Use of Warmer

1. Connect the unit to the mains. Switch it on.
2. Select manual mode.
3. Select heater output to 100% for sometime to allow quick pre-warming of the bassinet covered with linen.
4. Select servo mode.
5. Select the desired set temperature of baby as 36.5°C.
6. Place baby on the bassinet.
7. Connect skin probe to the baby's abdomen with sticking tape.
8. If you want the manual mode to be used in the baby, select the desired heater output.
9. In manual mode, record baby's axillary temperature at 30 minutes and then 2 hourly.
10. Respond to alarm immediately. Identify the fault and rectify it.

Application of Skin Probe

Do's

1. Prepare the skin using an alcohol/spirit swab to ensure good adhesion to the skin.
2. Apply probe over the right hypochondrium area in the supine position.
3. Apply probe to the flank in the prone position.
4. Check sensor probe regularly so as to ensue that it is in place. Ensure that skin probe is free of contact with bed.
5. Cover probe with a reflective cover pad, if available (foil covered foam adhesive pad).

Don'ts

1. Do not apply to bruised skin.
2. Do not apply clear plastic dressings over probe.
3. Do not use fingernails to remove skin surface probes.
4. Do not reuse disposable probes.

Use of Cling Wrap to Decrease Insensible Water Losses

Use of cling wrap (transparent polythene used for covering fruits or vegetable for storage) over the baby, tied across with the panels of warmer has been shown to reduce insensible water losses and result in better thermal control for VLBW (<1.5 kg) babies.

Potential Pitfalls of Servo-controlled Warmer

In the event of displaced probe from baby's abdominal skin, overheating of the baby will occur because the skin probe depicts air temperature and heater output keeps on increasing till probe temperature matches control temperature. In servo mode repeated activation of alarm will occur when baby develops fever. In this situation, one should shift to manual mode with least heater output.

Useful Tips for Use of Radiant Warmers

- Do not use the warmer in a cold room. It works best when the environmental temperature is above 20°C.
- Keeping the warmer where there is lot of air currents reduces its efficiency.
- The warmer must be pre-warmed around 20 minutes before the arrival of the baby or till the set temperature is reached with less than 50% of total heater output.
- While using the manual mode in a warmer without a temperature display, record the baby's temperature regularly, preferably 2 hourly.
- Train junior doctors and nurses about the proper use of servo and manual modes.
- The manual mode is used for initial preparation of bed for the baby; or when rapid warming of a severely hypothermic baby has to be done. However, this may be hazardous as babies may become overheated. Except in the continuous presence of a nurse who is watching the skin temperature, it is preferable to use the skin probe with the warmer on servo mode.

Disinfection

When the equipment is in use, all approachable external surfaces should be cleaned daily with an antiseptic solution like 2% bacillocid or glutaraldehyde. *Spirit or other organic solvents must not be used to clean the glass side panels or display panel.* For disinfection of reusable probe, isopropyl alcohol swab should be used.

Every seventh day, after shifting the baby to another cot, the used equipment should be cleaned thoroughly, first by light detergent solution and then by antiseptic solution. All detachable assemblies, are to be treated similarly.

Maintenance

Ongoing maintenance is the key to increase the mean time between failures. The hospital biomedical engineer must regularly check equipment but the authorized company engineer must be called for preventive checks and major breakdowns. The control and power units should be calibrated every 4–6 months and thorough servicing should be done annually. Temperature calibration should ensure sensitivity to ± 0.5° of the set value.

Costs, Models and Dealers

Costs may vary according to the make and are listed in Table 2.1.

TABLE 2.1: Available models and cost (*in thousand of rupees*)

Equipment	Indian	Imported
1. Radiant warmer		
• Manual controlled	15–35	50–60
• Servocontrolled	25–70	150–300

(*Costs vary widely with additional features*)

Manufacturers

Indian (approximate cost of radiant warmer ₹ 20,000 to 40,000/-), Shreeyash Electromedical (Pune), Zeal Medical (Mumbai), Meditrin (Mumbai), Bird Meditech (Mumbai), Medicaid (Delhi), Mediserve (Delhi), Mech Tech Lab (Pune), Nice Neotech (Chennai), Lectromedik Pvt. Ltd. (Bangalore), Medilek (Karnatka), Technomed Pvt Ltd. (UP), Doctor's Medical equipment, Ronak, Medical equipment India, Phoenix Medical systems (The CIC101, NWS 102, The NWS 101), AVI Healthcare Pvt. Ltd., Ved Med Software and Trading Pvt. Ltd. (Fig. 2.2).

14 Neonatal Equipment

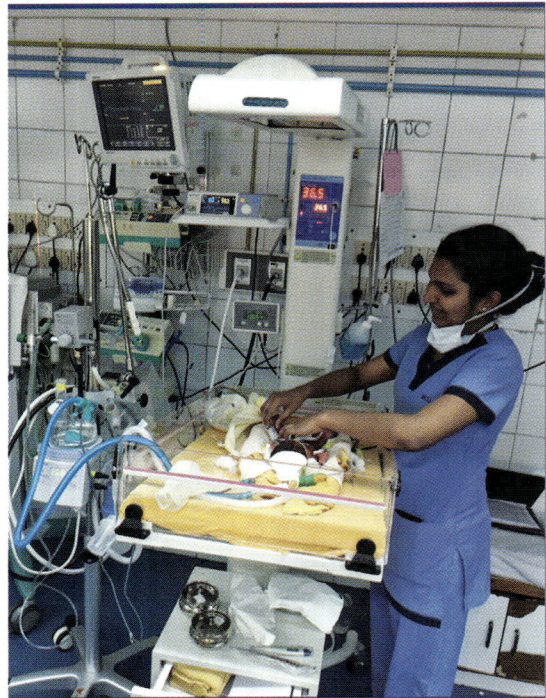

Fig. 2.2: Indigenous radiant warmer—staff taking care of sick premature baby

Dealers of Imported Brands (Fig. 2.3)

Principals	Dealers (Brand)
Ameda (Switzerland)	Medisphere
Datex-Ohmeda GE	Wipro GE Healthcare
Fisher-Paykel	Fisher and Paykel (Cosy Cot-IW-931, IW-932, 933)
Premicare (USA)	Moola Tools Pvt. Ltd.
AG (Germany)	GE Wipro
Heraeus (Germany)	Medex
Atom (Japan)	Vishal Surgicals
Drager	Drager India (Air-Shields[R] Infant Care System)
Wipro GE Healthcare	GE Healthcare
Fanem (Brazil)	SMRA International (Telangana) Brands (Ampla 2085)
Ginevri (Italy)	Global Medical System (Alhena and Alhena plus)
Weyer Gmbh (Germany)	Rustagi Surgicals (Variotherm)

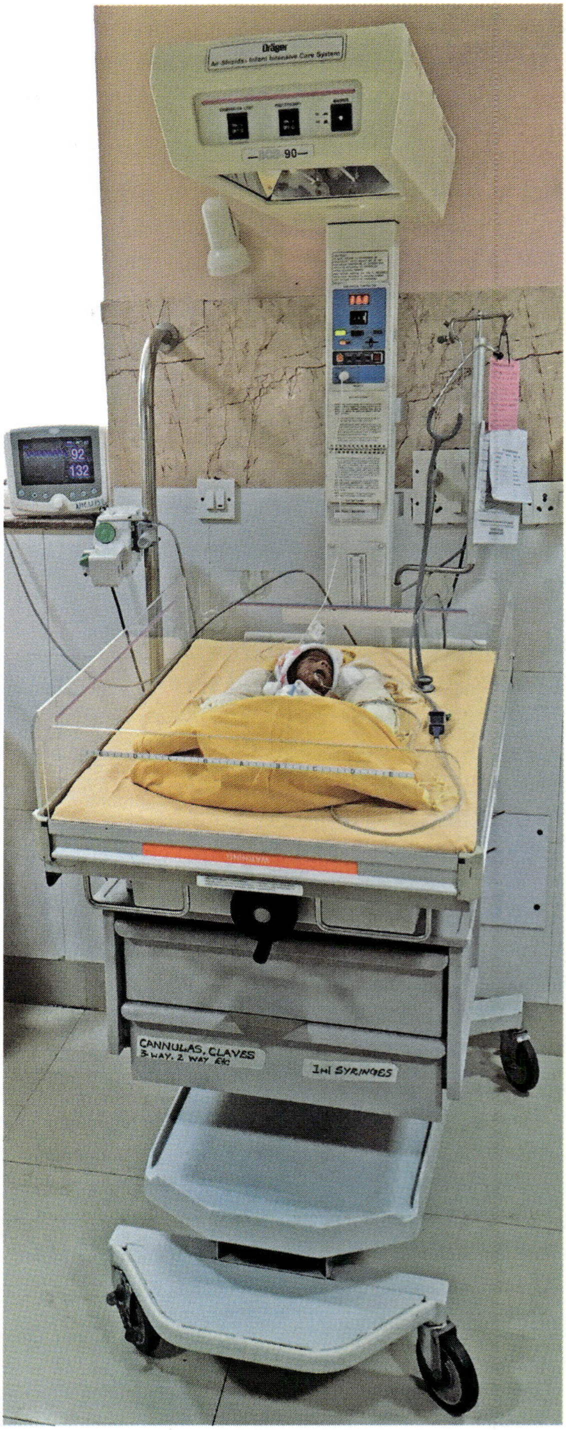

Fig. 2.3: Premature baby under an imported radiant warmer

Conclusion

The use of warmers is now firmly established in special care units. The sophistication of equipment has also reached a mature state. Familiarity with the equipment and the control system, proper use, cleaning, disinfection and daily maintenance are of paramount importance.

On the whole, Indian warmers have greatly improved in quality over the last decade. Their costs have also increased simultaneously. A variety of models from both Indian and foreign companies are available. In general, imported models still give longer trouble-free service and more accurate temperature control. However, the cost of a single imported unit can buy two to four Indian products. The bottom line is the quality of after-sales service, which often is equally unsatisfactory whether the companies deal with indigenous or imported equipment. It is best, therefore, to consult other colleagues using different models before purchasing one. Apart from quality and performance of a model, its cost, availability of spare parts and servicing facilities are also important considerations.

Frequently Asked Questions (FAQs)

Q. 1. What are indications for use of manual mode in a warmer?

Manual mode is used for following situations
1. When you are anticipating a new baby to be brought under warmer care. Keep warmer on with 100% heater output, once the baby is arrived shift to servo mode.
2. If baby is having fever, move to manual mode and make heater output minimum. If the baby continues to be in servo mode alarm will get activated time and again.
3. When rapid warming of hypothermic baby has to be undertaken do using manual mode.
4. In labor room when attending delivery, the manual mode alarms every 10 to 15 minutes indicating the bed is warm and ready. If alarm is silenced, again it will reactivate after another 10–15 minutes.

Q. 2. What tips should be followed for the use of open care system?

Tips for use
1. Connect the unit to the mains. Switch it on.
2. Select manual mode.
3. Select heater output to 100% for some time to allow quick pre-warming of the bassinet covered with linen.
4. Select servo mode.
5. Select the desired set temperature of baby as 36.5°C.
6. Place baby on the bassinet.
7. Connect skin probe to the baby's abdomen with sticking tape.
8. If you want the manual mode to be used in the baby, select the desired heater output.
9. In manual mode, record baby's axillary temperature at 30 minutes and then 2 hourly. Respond to alarm immediately. Identify the fault, rectify it.

Q. 3. What precautions should be followed for application of skin probes?
1. Do not apply to bruised skin.
2. Do not apply over clear plastic dressings.
3. Do not use fingernails to remove skin surface probes.
4. Do not reuse disposable probes.
5. Shield skin probes with reflective pad, if possible, under radiant warmer.
6. When using servo control mechanisms for environmental control, take intermittent temperatures at other sites to monitor effect.
7. Check sensor probe regularly so as to ensure that it is in place.

Q. 4. What is the technique for application of skin probes?
1. Prepare the skin using an alcohol/spirit swab to ensure good adhesion to the skin.
2. Apply probe over the right hypochondrium area in the supine position.

3. Apply probe to the flank in the prone position.
4. Ensure that skin probe is free of contact with bed.
5. Cover probe with a reflective cover pad (foil covered foam adhesive pad).

Q. 5. What are the advantages and disadvantages in using a warmer?

Advantages
1. Easy accessibility.
2. Easy to connect the tubes of ventilated baby and do procedures.
3. Better monitoring especially if baby has respiratory distress.
4. Less risk of infection as compared to closed incubator.
5. Can be used as resuscitation trolley in the labor room.

Disadvantages
1. More insensible water losses.
2. Not uniform heating as compared to closed system.
3. More risks of episodes of hypothermia.

Q. 6. Does one need to record baby temperature if baby is under radiant warmer?

Yes, the axillary temperature must be recorded and documented as per nursery policy. Compare this with the depicted display temperature. This gives the opportunity to be sure that the warmer is working all right and baby is not over or under heated.

Q. 7. If baby is having higher temperature, how one can be sure that this is due to illness in the baby or overheating by warmer?

If baby is having fever, examine the baby carefully. A baby who is overheated due to warmer will have his skin red flushed, the temperature of sole/palms will be warm to touch by dorsum of hand in addition to warm abdomen. Malfunction of equipment—probe getting disconnected or keeping baby in manual mode with high heater output can explain this situation. On the other hand fever due to illness will result in warm abdomen to touch but palms/soles will be cold to touch (gradient between abdomen and palms/soles temperature). In addition, clinical examination will reveal features that point towards sepsis in the baby.

Chapter 3

Incubators

Lack of attention to thermoregulation continues to be a cause of unnecessary deaths in the neonatal population. Maintaining a stable body temperature is essential to ensure optimal growth. If temperature is maintained, caloric expenditure and oxygen consumption is minimal. Newborn babies, in particular the preterm and the low birth weight are exquisitely predisposed to hypothermia. No other equipment is identified more with the special care of newborn babies than the warming devices, namely incubators and radiant warmers.

THE PHYSIOLOGY OF THERMOREGULATION IN NEWBORN

Heat loss from a body occurs by conduction, convection, radiation or evaporation. *Conductive* losses are due to difference in temperature between the baby and the surface on which he/she lies. *Convective* losses are dependent on air currents around the baby. *Evaporative* losses are due to latent heat of evaporation and depend on how wet the baby is, the thickness of the skin and humidity of environment? *Radiation* heat loss depends on the temperature gradient between two objects facing each other.

Compared to adults, newborn babies are at a handicap due to a relatively large surface area per unit mass and reduced heat generation by means of shivering. Preterm babies lie in an extended position, unlike the flexed posture of term newborns and hence have larger exposed surface area. This compromises their ability for heat conservation. They also have insufficient brown fat, less subcutaneous fat and inefficient control of dermal regulation of blood flow, which contribute towards lower thermal insulation.

For a particular gestation and postnatal age there exists an optimal environmental temperature at which oxygen consumption of the baby is the lowest. Temperatures both above and below this range result in increased metabolic rate. This ambient temperature range is called the thermoneutral temperature.

INCUBATORS

Incubators reduce convective and radiation heat losses by reducing exposure to air currents and by providing a warm environment. Evaporative losses are minimized by maintaining high humidity in the incubator. Radiation losses are curtailed by the hood or canopy on the baby or by using double-walled incubators. Radiant warmers provide an intense source of radiation heat energy. They also reduce the conductive losses by providing a warm micro-environment surrounding the baby.

The incubator is a closed system with a heating element underneath and a transparent hood or canopy around the baby tray. Air or an air-oxygen mixture is sucked in through a microfilter, streamed over the heating element and the humidifier using a quiet fan. The warm humidified air is then circulated through the hood to attain a uniform temperature within. A low rate of air circulation, ideally not more than 20–30 liter per minute, minimizes convective heat losses due to fast currents around the baby. Noise level within the incubator is kept below 60 dB to avoid deleterious effects on hearing.

The hood or canopy (acrylic/plexiglass/fiberglass) in incubators should provide clear vision for several years. Hood design may be single or double-walled. Three controlled trials have compared double-walled with single-walled incubators. Two of these trials showed a reduction in oxygen consumption in the double-walled incubator, while one showed no definite advantage of a double-walled incubator. Access holes, called iris ports, are often elbow operated to open and close. The under deck area and conditioning chamber should be corrosive resistant, moulded and easy to clean. Other components include air inlet, filter, fan or blower system, humidity tray, and baby tray with the mattress, an inlet for oxygen and IV tubings and baby and air temperature probes. Good quality equipment should have smooth surfaces and be free of corners and crevices. Additional optional features which may be attached to the incubator include an intravenous stand, weighing scale, timer, tilt facility, battery back up, oxygen analyzer, resuscitator, vital signs monitor, phototherapy unit, oxygen flow meter, suction, ventilator, etc.

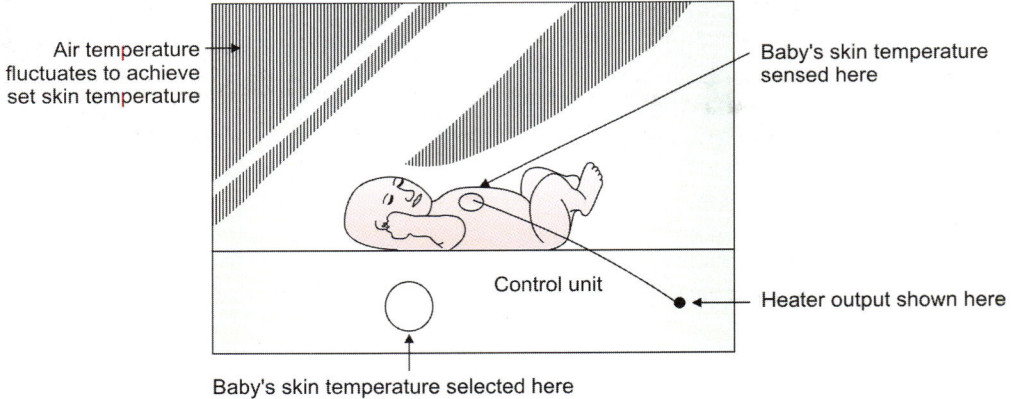

Fig. 3.1: Incubator—servo-controlled mode

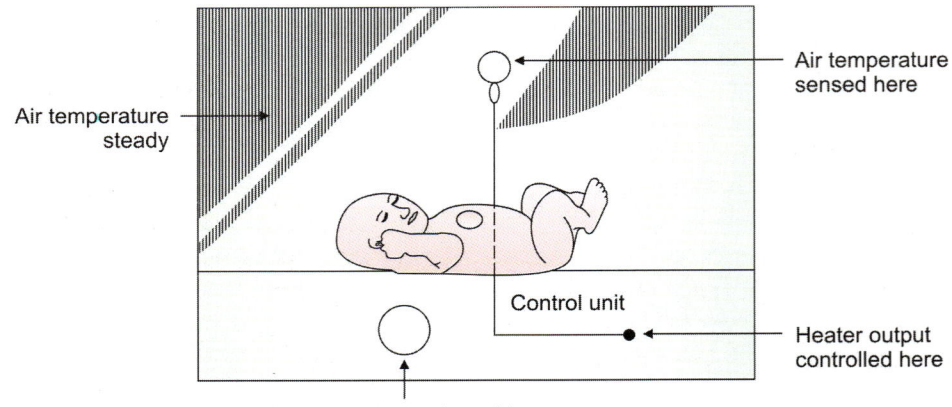

Fig. 3.2: Incubator—air-controlled mode

Possible functions of a neonatal incubator can be summarized as:
- *Protection* from cold temperature, infection, noise, drafts and excess handling: Incubators may be described as bassinets enclosed in plastic, with climate control equipment designed to keep them warm and limit their exposure to germs.
- *Maintaining* fluid balance by providing fluid and keeping a high air humidity to prevent a too great loss from skin and respiratory evaporation.
- *Oxygenation*, through oxygen supplementation.
- *Observation*: Modern neonatal intensive care involves sophisticated measurement of temperature, respiration, cardiac function, oxygenation, and brain activity.

TEMPERATURE CONTROL MODES

There are generally two modes which are used. In the *air mode*, desired temperature around the baby is set and the heater output adjusts itself to maintain this. The appropriate set temperature is decided by using thermoneutral temperature charts (Table 3.1) as applicable to an individual baby based on gestation and postnatal age. The air temperature sensing probe should be placed near the baby and care should be taken that it is not displaced or covered.

In the *servo-controlled mode* or the *skin temperature controlled mode*, the desired skin

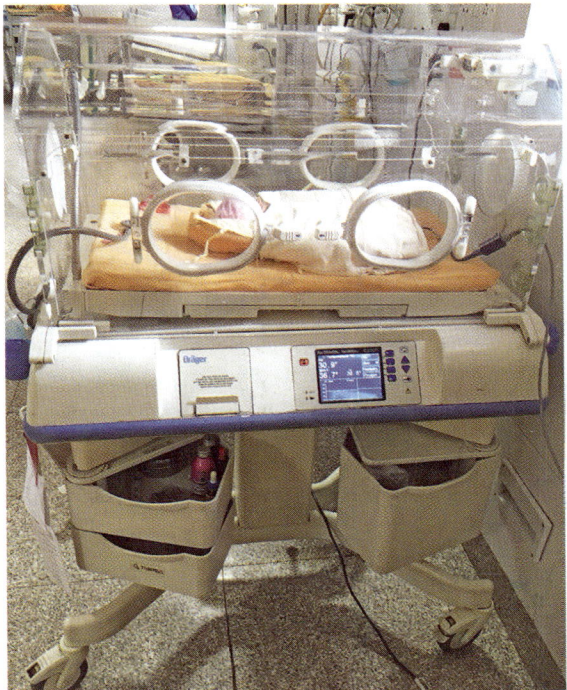

Fig. 3.3: Servo-controlled double-walled intensive care incubator with display of air, control and baby's temperature

temperature is set to 36.5°C (Fig. 3.3). The baby's temperature is monitored by a skin probe, fixed firmly using a tape on the anterior abdominal wall. The feedback system modifies heater output to keep the baby temperature constant. The advantage of this method is a reduced need for close monitoring of the newborn temperature. However, this method may mask hypo- or hyperthermia in the baby and may cause accidental overheating due to displacement of the skin probe.

Manually controlled fixed or variable heater output radiant warmers are also available and are useful in babies who are not too small or sick and where precise control of temperature is not crucial.

Humidification

Humidification of incubators reduces insensible water losses. It is achieved by filling the humidification tray with ½ liter of distilled

TABLE 3.1: Neutral range of environmental temperature (C) for low birth weight babies

Age	Birth weight (gm)		
	<1200	1200–1500	1501–2500
1st day	35.0 ± 0.5	34.3 ± 0.5	33.4 ± 1.0
2nd day	34.5 0.5	33.7 ± 0.5	32.7 ± 1.0
3rd day	34.0 ± 0.5	33.5 ± 0.5	33.0 ± 1.0
≥4th day	33.5 ± 0.5	32.8 ± 0.5	32.2 ± 1.0

(Adapted from Standards and Recommendations for Hospital Care of Newborn Infants, Ed 5; Evanston, Illinois, 1971; American Academy of Pediatrics)

water with 1–2 ml of glacial acetic acid or vinegar to prevent bacterial colonization. In some incubators, water soaked sponges are used. The water must be changed daily. Generally, in practice, humidification is employed only for babies less than 1 kg who have high evaporative losses. The main risk of humidification is the fear of Pseudomonas colonization in the water.

Disinfection

When the equipment is in use, all approachable internal and external surfaces should be cleaned daily with soap water or antiseptic. *Spirit or other organic solvents must not be used to clean the incubator hood or panel.*

Every seventh day, after shifting the baby to another clean incubator, the used equipment should be cleaned thoroughly, first by light detergent solution and then by antiseptic solution (e.g. 20% bactishield H_2O_2 + $AgNO_3$). All detachable assemblies, especially from the under deck area, are to be treated similarly.

After drying, the parts are reassembled and sterilized using a vaporizing agent and/or fumigation. Adding 50 ml of formalin to 50 ml of distilled water in humidity tank and plugging it for 4 hours leads to fumigation of the incubator. Glutaraldehyde (2%) or 20% bactishield (11% w/v hydrogen peroxide and silver nitrate 0.01% w/v) is a good alternative. After fumigation it should be thoroughly aired. The sleeves of the access windows must preferably be changed daily and cleaned.

Maintenance

Ongoing maintenance is the key to increase the mean time between failures. The hospital biomedical engineer must regularly check equipment but the authorized company engineer must be called for preventive checks and major breakdowns. Air filters generally require change every 3 months or if they are visibly dirty. A clogged filter reduces the oxygen entry and promotes carbon dioxide build up in incubator. The control and power units should be calibrated every 4–6 months and thorough servicing should be done annually. Temperature calibration should ensure sensitivity to ±0.5° of the set value.

Useful Tip for Use of Incubators

- Double-walled incubators or dome shields may be used if environmental temperatures are low.
- Room temperature between 21° and 28°C is ideal.
- Positioning the incubator parallel to the wall should be avoided because it hampers air circulation.
- Avoid humidification unless insensible water losses are high as in a baby weighing <1 kg. Water must be changed daily if humidification is being used.
- Whenever front panel is opened in the servo mode, one should change over to air control mode.
- Do not use alcohol to clean the canopy, this will make the Perspex™ foggy.

Radiant Warmers Versus Incubators

Radiant warmers score over incubators because of superb accessibility to the baby and ease in performance of procedures. This may however be a mixed blessing because it may encourage unnecessary handling. Its major drawback is the excessive evaporative heat losses. This is not tolerated by the extremely low birth weight baby, whose skin is thin, resulting in enormous insensible water losses. The effectiveness of radiant warmer is compromised if it is placed where there is strong draught as this causes excessive convective heat losses. Major differences between the two devices are summarized in Table 3.2.

Costs, Models and Dealers

Costs may vary according to the make and are listed in Table 3.3.

TABLE 3.2: Radiant warmer versus incubator (major differences)

Radiant warmer	Incubator
1. Open system	1. Closed system
2. Overhead heater with reflector to focus radiant heat	2. Under deck heater and fan to circulate warm filtered air
3. Easy access to baby	3. Baby less approachable
4. Increased insensible water losses (IWLs about 20–25 ml/kg)	4. IWL is less and can be further reduced with humidification
5. Easy to clean and maintain	5. Difficult to clean and maintain
6. Lower risk of infection unless infant excessively handled	6. Risk of infection increases, if not adequately cleaned and disinfected
7. Less cost	7. Costlier

TABLE 3.3: Available models and cost (*in thousand of rupees*)

Equipment	Indian	Imported
1. Incubator		
• Air controlled	20–50	150–400
• Skin controlled	80–100	250–500

(Costs vary widely with additional features)

Manufacturers

Indian: Shreeyash Electromedical (Pune), Zeal Medical (Mumbai), Meditrin (Mumbai), Phoenix (Chennai), Nice Neotech (Chennai), Medicaid (Delhi), Mediserve (Delhi), Mech Tech Lab (Pune), Lectromedik Pvt. Ltd. (Bangalore), Medilek Instrument Vcare system, Medical equipment, SS Technomed, Perfect technologies, Indian surgical industries, Universe surgical equipment, Kriticare India, Medicaid, Novel health care product, Atico Medical Pvt. Ltd. (infant baby incubator AM-1408).

Dealers of Imported Brands

Newer dual mode incubator and radiant warmer for seamless conversion of both modes, e.g. Giraffe Omheda (Price ₹ 14,00,000).

Principals	Dealers (Brand)
Ameda (Switzerland)	Medisphere (Medica 2015)
Ginevri (Italy)	Global Medical System (Polytrend)
Fisher–Paykel (NZ)	Fisher and Paykel (Bangalore)
Weyer Gmbh (Germany)	Rustagi Surgicals (Thermocare, Thermocare vita)
Heraeus (Germany)	Medex
Atom (Japan)	Vishal Surgicals (Transcapsule V-808)
Datex Ohmeda (USA)	Wipro GE Healthcare
Draeger	Draeger (Isolette 2000, 8000)
Fanem (Brazil)	Cardiocare (Vision advanced 2286, 1186A, 1186B)
GE Healthcare USA	Wipro GE Healthcare (Lullaby XP)
Draeger	HL Medical (T1500)
Natus Medical (USA)	Rohanica (natal care)

Conclusion

The use of warmers and incubators is now firmly established in special care units. The sophistication of equipment has also reached a mature state. Familiarity with the equipment and the control system, proper use, cleaning, disinfection and daily maintenance are of paramount importance.

On the whole, Indian incubators have greatly improved in quality over the last 10–15 years but still more to be done. A variety of models from both Indian and foreign companies are available. In general, imported models still give longer trouble-free service and more accurate temperature control. The bottom line is the quality of after-sales service, which often is equally unsatisfactory whether the companies deal with indigenous or imported equipment. It is best, therefore, to consult other colleagues using different models before purchasing one. Apart from quality and performance of a model, its cost, availability of spare parts and servicing facilities are also important considerations.

Frequently Asked Questions (FAQs)

Q 1. What are the indications for use of incubators?

Incubators may be used for the following indications

1. For neonates born <35 wk and/or birth weight <1800 g.
2. For humidification, specially for extremely low birth weight babies <1 kg.
3. For isolating an infected baby to achieve barrier nursing.
4. For providing oxygen, depending on oxygen flow rate, ambient oxygen concentration can be maintained.
5. For transporting babies (using transport incubator).
6. For providing stimulation to an apneic baby, if incubator has rocking bed attachment.
7. For use at extremely low ambient temperatures (<20°C) or when there are lot of convective currents where a radiant warmer fails to work.

Q. 2. What precautions should be followed for application of skin probes?

1. Do not apply to bruised skin.
2. Do not apply over clear plastic dressings.
3. Do not use fingernails to remove skin surface probes.
4. Do not reuse disposable probes.
5. Shield skin probes with reflective pad, if possible, under radiant warmer.
6. When using servo control mechanisms for environmental control, take intermittent temperatures at other sites to monitor effect.
7. Check sensor probe regularly so as to ensue that it is in place.

Q. 3. What is the technique for application of skin probes?

1. Prepare the skin using an alcohol/spirit swab to ensure good adhesion to the skin.
2. Apply probe over the right hypochondrium area in the supine position.
3. Apply probe to the flank in the prone position.
4. Ensure that skin probe is free of contact with bed.
5. Cover probe with a reflective cover pad (foil covered foam adhesive pad).

Q. 4. What are the types of probes available for recording temperature in the newborn?

Basically these are of two types

1. *Thermistor probe* (most widely used). The thermistor is a resistive component having a high negative coefficient of resistance, so that its resistance decreases proportionately as temperature rises. As the resistance of the thermistor changes the electrical current flowing through the probe changes proportionately. The level of current detected by the electronic monitor is converted to thermal units.
2. *Thermocouple probes*: Thermocouple probe is a very small bead made up of the junction of two dissimilar metals. The bead generates a very small voltage proportional to temperature. The voltage generated by the bead is measured by the monitor and converted to thermal units.

Q. 5. What are the potential pitfalls of servo-controlled heating devices?

In the event of displaced probe from baby's abdominal skin, overheating of the baby will occur because the skin probe depicts air temperature and heater output keeps on increasing till probe temperature matches control temperature. This means that air temperature matches control temperature which is not physiological. Servo mode will result in repeated activation of alarm in situations where baby develops fever. In this situation, one should shift to manual mode with least heater output.

Q. 6. Describe generation of greenhouse effect with use of incubator.

An incubator exposed to direct sunlight will let short wave radiation pass through the canopy ; while longer wave lengths radiation directly heat the walls. This in turn radiates heat to baby leading to overwarming.

Q. 7. Enumerate fundamental characteristics of incubators available in the market.

The following types of incubators are available in the market:
1. Air temperature control.
2. Air temperature and skin control.
3. Air temperature, skin control, humidity level and oxygen concentration monitoring.
4. Air temperature, skin control, humidity level and oxygen concentration (servo control).

Q. 8. What are the criteria used for choosing the type of incubator?

The choice of an incubator should be made according to the level of neonatal care for which it will be used:
1. Minimum care
2. Sub-intensive care
3. Intensive care
4. Super-intensive care.

Q. 9. Do neonates weighing more than 1500 g need incubator care? If yes, then what is the best mode?

For newborns weighing more than 1500 g, it is generally sufficient to create a humidified micro-climate with a constant temperature and if necessary enriched with oxygen, at levels which can be determined easily by the baby's reactions. To provide this minimum care, it is sufficient to heat and thermoregulate the air in circulation (air mode) at 2°/4°C less than the body temperature required.

Q. 10. What type of incubator should be used for newborn weighing less than 1500 g?

For newborns weighing less than 1500 g and who have a very critical metabolism, thermoregulation of the air must be adapted to the needs of the patient. Therefore the thermoregulation sensor should be placed directly on the patient's skin instead of in air. In this mode (*skin mode*) the air inside the incubator will heat up until the baby has reached the required skin temperature. In this way the temperature of the air in circulation might rise to the set skin temperature. This happens when the newborn is very small and not able to provide his own body heat. As the baby matures, the temperature of the air in circulation and the heater output tends to decrease. When the baby is able to thermoregulate autonomously, the air temperature inside the hood will be the same as the room temperature, the heater output is nil and the skin temperature of the newborn is approximately 36°C.

Q. 11. What is an intensive care incubator?

Newborns with extremely low weight (<1 kg) and in critical condition need a microclimate in which the humidity and the oxygen concentration are controlled very precisely and at customized levels.

Obviously, for these patients, the incubator requires a control panel, which allows servo-controlled adjustment of the humidity percentage and of the oxygen concentration inside the hood, as well as the servo-controlled temperature adjustment (air and skin).

When a high oxygen concentration is required, varying from 50 to 100%, it is better to use a head box inside the incubator hood. The use of head box guarantees, with very low flow, oxygen enriched environment for the baby without altering the microclimate. Without the head box the high oxygen flow required to reach the desired concentration, would inevitably affect the humidity and temperature values. The head box should be designed with an automatic oxygen-sensor detection system to avoid oxygen enrichment when the sensor is not positioned or not properly positioned in the head box.

For tiny neonates (<1 kg) it is preferable to use equipment with a humidification system which provides "truly sterile" humidity. When the doors are fully opened it is generally better to use incubators which allow the newborn to be heated "statically", in other words, without any increase in ventilation around the newborn.

Q. 12. How does one decide that the baby no longer requires incubator care?

If in servo/skin mode the baby is able to maintain temperature with minimal (<25%) or no requirement of heater output for at least 24 hours, or in an air mode at 30°C, the baby can maintain normal temperature for 1 or 2 days, the baby can be taken out of the incubator. For ELBW babies decision is based on availability of radiant warmers in the unit and often this period may vary from 10 to 20 days.

Q. 13. How does heat loss pattern differs in a double-wall incubator?

In a single wall incubator, a newborn baby loses 2% heat by conduction, 16% by evaporation, 24% by convection and 58% by radiation. In a double-wall incubator, the radiation heat losses are reduced to 29% while the other losses remain the same.

Q. 14. Why is a double-wall incubator better when compared to a single wall?

In a single wall incubator, the difference of baby temperature (36.5°C) to incubator wall temperature (30.5°C) is nearly 6°C when the incubator is kept at ambient temperature of 25°C and air temperature inside the incubator being 36.5°C. This increases radiant loss from the baby tremendously. While in a double-wall incubator, under similar conditions inner incubator wall temperature is nearly 33.5°C, reducing the gradient from baby to wall to only 3°C. This reduces radiant heat losses from the baby by nearly 50%.

Q. 15. How are the heat losses minimized from a baby inside the incubator?

Raising the temperature of the inner wall of the incubator dramatically reduces the radiant heat losses, while raising the level of humidity using the servo humidifier will reduce the evaporative heat losses. Low air velocity over the baby will minimize convective and evaporative heat losses.

Q. 16. Do babies need to be clothed when under a radiant warmer or in an incubator?

Baby is clothed if nursed inside the incubator. On the other hand, if the baby is sick and nursed under a radiant warmer, he is kept naked except for covering the head, feet and hands. Even under the radiant warmer, the baby is clothed once he is stable and growing.

Chapter 4

Transport Incubators

Regionalization of perinatal care would necessarily lead to transport of newborn infants to and fro from one center to another. Transport of babies is also required for carrying babies within the hospital from one place to the other. In order to ensure safe transport, one would need a co-ordinated service and a transport incubator. The best transport incubator is the uterus but *in utero* transport may sometimes not be possible. This is when a transport incubator along with a transport service becomes essential.

TRANSPORT INCUBATOR

A *transport incubator* is an incubator and the most necessary neonatal instruments in a transportable format, and is used when a sick or premature baby is moved, e.g. from one hospital to another, as from a community hospital to a larger medical center with a proper neonatal intensive care unit. It usually has a miniature ventilator, cardiorespiratory monitor, IV pump, pulse oximeter, and oxygen supply built into its frame. It differs from a standard incubator in having an alternative source of power or heat.

A transport incubator has following essential functions

a. Ability to provide additional warmth
b. Facilities for observation and monitoring
c. Facilities for medicating/intervening if necessary
d. Portability, ruggedness (good wheels, stand for IV, etc.)
e. Ease of attaching additional equipment if necessary
f. Easy cleaning and disinfection.

Types of Transport Incubators

Three types of transport incubators are described based on the type of heat source:
1. Transport incubator with no heat source
2. Warm water transport incubator
3. Electricity operated transport incubator.

Transport Incubator with No Heat Source

These are low cost indigenously manufactured thermocole boxes. Holes are cut in the box to ensure observation and air circulation. The newborn is kept well wrapped using blister bubble wrap silver foil pre-warmed cotton or even blankets for prevention of heat loss. Advantages of this type of incubator include local fabrication, low cost, ease of use and wide availability. Disadvantages are that very sick infants are difficult to manage and close observation is sometimes difficult when they are being transported. Such incubators are in use in many parts of our country and reports of successful usage in primary care settings, remote and inaccessible areas have been published. Cleaning and disinfecting of such incubators is sometimes difficult, often it is easier to find a new thermocole box and build another one.

Warm Water Transport Incubator

Warm water is an efficient heat sink, but has the hazard of causing scalds or burns if the infant comes in contact or is very close to the water bottle. It has been used because of the wide availability of hot water and hot water bottles. The hot water bottles are placed in polythene casings to prevent leaks and wrapped in multiple layers of cloth to prevent contact burns. The bottles can then be placed in thermocole boxes and used as warm water transport incubators.

An incubator using hot water as heat source for heating air has been designed and can be fabricated indigenously. This indigenous incubator has a large metal can fitted on one wall of incubator (Fig. 4.1). When the can is filled with water, circulating air currents distribute the warmth to rest of the incubator. The advantage is low cost, low maintenance and ease of use. The disadvantage is that once the can is filled with hot water, the incubator remains warm for only about 4 hours. The incubator is very heavy when water is filled in. The simplicity of its design and low cost make it an ideal incubator for short distance transports. The main caution as with any hot water incubator is to ensure that there is no spillage of water. The design ensures that even if there is a little spill it will probably not come in contact with the baby, hence the risk of scalds is small.

Electrically Operated Transport Incubators

These are very much like conventional incubators but are portable, have an inbuilt battery as a source of power and can have provision to run on power supply from an ambulance. Many imported brands are currently available while some indigenous manufacturers are willing to fabricate one if given the specifications. The electrically operated transport incubator functions like the usual incubator in having temperature setting, persplex canopy for observation, portholes for access to the baby. It differs in being smaller in size, lighter in weight, relatively resistant to vibrations, and having provision for attaching monitoring equipment, pumps, etc. (fully loaded). Some brands come with built in flow meters, provision for air and oxygen cylinders, observation lights, temperature probes, etc. Some brands have facility for attaching other instrument separately (add on equipment).

TRANSPORT—CONDITIONS DURING USAGE

The work environment of transport differs from a nursery setting in the following significant ways excessive noise, vibration, improper lighting, variable ambient temperature, limited availability of drugs, and supply, limited support services.

The transport incubator has to function in all these adverse circumstances and has to be user-friendly to the staff overseeing the transport. Since the incubator needs to compensate for these problems it needs regular preventive servicing and maintenance.

- *Preventive maintenance*: Premature malfunction is a very common feature of equipment used in transport. Hence, preventive maintenance and pre-transport self checks should be part of the purchase deal for any transport incubator.
- *Electronic monitoring*: Physical examination of the neonate during transport is difficult, thus the transport incubator must have provision for providing for electronic monitoring of vital parameters.

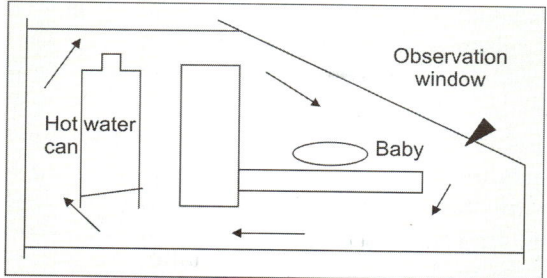

Fig. 4.1: Warm water transport incubator

ADD-ON EQUIPMENT

Transport incubators manufactured abroad come with a host of built in equipment. These include ventilator, vital sign monitor, pulse oximeter, capnograph, transcutaneous monitors, and infusion pump. All these pieces of equipment need to have battery backups as power failures are common. The need for this equipment is governed by many factors. Firstly, it is virtually impossible to do a physical examination on a baby while on the move. Secondly, vibrations, lack of space and inclement environment makes handling of the baby very difficult while being transported. Thirdly, equipment which has been added on must function well in adverse circumstances. For example, we have found that standard ECG monitors do not function during ambulance transport in Bengaluru as the vibrations cause multiple artifacts. Thus, it is preferable to "road test" the equipment prior to purchase.

Essential Equipment to go along with Transport Incubator (Fig. 4.2)

Oxygen cylinder, flow meter, tubing, masks, head box, self-inflating resuscitators and manual/electrical suction devices are essential equipment to take along with every transport incubator. A good focusing examination light is of added advantage, particularly when ambient light conditions are likely to vary, and if night transports are contemplated. There is a long list of medicines and supplies that a transport team needs to carry along with the incubator. Many such lists are available in standard textbooks on the subject.

Usefulness of a Transport Incubator

Units catering to predominantly extramural (out born) babies certainly will benefit from a transport incubator. Units where the delivery room is far away from the nursery also require a transport incubator. Transport incubators also have a role in reverse transfer from tertiary to secondary or primary centers. In all these settings a transport incubator helps reduce thermal stress and promotes better outcomes as healthier babies reach the NICU. The type of incubator to acquire depends on:

a. The distance of transport
b. The type of illnesses and degree of severity of illness in babies catered to small, high risk infants require closer monitoring and hence a more sophisticated incubator and transport service is required. Surgical neonates have special requirements which need looking into if the unit has a large number of surgical referrals.
c. Number of transports envisaged per day. Greater the number more rugged and easy to clean should the equipment be.
d. Time and distance of transport—longer transports need longer battery backups as well as greater amount of supplies.
e. Affordability.

Shortcomings and Complications

Transport incubators need to be light and should need only two people to carry it. The foreign hi-tech incubators are heavy and sometimes seven to eight people are needed to carry them on to ambulances. Battery failure is common if maintenance is lax, and hypothermia may result from technical rather than medical reasons. Power fluctuations or power failure can easily destroy delicate electrical

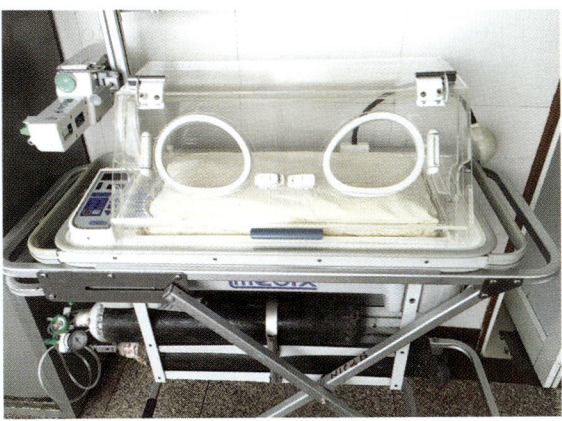

Fig. 4.2: Light weight transport incubator with portable ventilator, runs on chargeable battery

circuits. The transport incubator needs to be hardy enough to withstand power fluctuations in more than one hospital and ambulance. Uneven roads and erratic traffic, means that a baby must be strapped or "packed tightly" into an incubator, otherwise injury may occur. Warm water based incubators must particularly be checked to ensure burns/scalds are not caused. Hyperthermia or hypothermia has resulted because of incubator malfunction and careful monitoring during transport (particularly long ones) needs to be done. Ultimately, it is the quality, dedication, training and maintenance capabilities of the transport team which determines the outcome of the transport rather than the equipment or its brand.

Frequently Asked Questions (FAQs)

Q. 1. How to go about buying a transport incubator?

Following will decide the type and need of transport incubator for your unit

1. What sort of babies (medical surgical; weight, gestation)? Are you catering to or are you likely cater to?
2. How often is a transport service likely to be required?
3. How much money can you afford?
4. What is the distance you are likely to transport babies?
5. How good is your biomedical team in terms of maintaining equipment?

Look for

1. Ruggedness
2. Service and maintenance record of the company selling you the equipment
3. Road-test before purchasing.

Q. 2. What are the ideal specifications for a transport incubator?

A transport incubator should be light weight, portable, made up of material which can be easily cleaned. It should run on mains or battery. The incubator should maintain air temperature within the desired range and must have the facility of providing oxygen and suction. The baby under transport should be visible through the transport canopy. Sophisticated transport incubator will have additional add-on's like ventilator, vital sign monitor, etc. and facility for keeping baby in servo mode.

Common imported transport incubators available in the market

Make	Dealer's name	Unit cost (₹)
Indian	Meditrin Instruments Mediserve Nice Neotech Med Sys Pvt. Ltd. Zeal Med Pvt. Ltd. Phoenix Med Sys Pvt. Ltd.	1,00,000 to 1,50,000

Imported transport incubators

Draeger	Draeger (Isolette 1500)
Fanem (Brazil)	Cardiocare (IT-158 TS)
Draeger	HL Medical (T1500, Globe Trotter)
Natus Medical (USA)	Rohanica (Tr-200, Tr-306)
Arishields Vickers	Wipro GE Healthcare
Atom V 808	Vishal Surgicals
Ginevri (Italy)	Global Medical System (baby shuttle)
Amertans	Medisphere
Advanced International (Miami, Fl)	Advanced instrumentations, India (A3158)

Specifications for an electrically operated transport incubator
- Dimensions (approx) width 50 cm
- Weight, height 100 cm, length 150 cm, not more than 50 kg
- Electronic temperature regulation and control
- Temperature probe
- Internal lighting
- *Rechargeable battery*—able to run incubator, lighting, suction unit, monitors, infusion pumps, ventilator, etc.
- Power system capable of charging battery from mains/ambulance power source
- Air/oxygen blender and flow meter
- Oxygen and air tank/compressor
- Straps for baby

- Fitting rail for accessories
- Operation/service manual
- Suction machine and suction device
- Preventive maintenance checklist
- Service contract, annual maintenance contract as for any other equipment.

Q. 3. What equipment, medicines and nursing supplies need to accompany the transport incubator?

Medicines
1. Adrenaline — 4 ampoules
2. a. 10% dextrose — 1 bottle
   b. Isotonic saline — 1 bottle
   c. Pediatric electrolyte solution — 1 bottle
3. Heparinized saline — 5 ampoules
4. Dopamine — 1 ampoule
5. Sodium bicarbonate — 1 ampoule
6. Vitamin K — 1 ampoule
7. Phenobarbitone — 1 vial
8. Naloxone — 2 vials
9. Midazolam — 1 vial

Equipment
1. Ambu bag + reservoir + oxygen tubing
2. Face mask — 2
3. Infant laryngoscope with 0 and 1 blades — 1 each
4. Spare AA batteries — 2
5. Sticking plaster — 2 rolls
6. Scissors — 1
7. Ventilator set — 1
8. Mucus sucker — 2
9. Endotracheal tubes — 3
10. Infusion pumps — 1
11. Pulse oximeter — 1

Nursing supplies
- 24 G neoflon — 2
- 22 G venflon — 1
- 23 G butterfly — 2
- 50 ml syringe — 2
- 20 ml syringe — 2
- 10 ml syringe — 4
- 5 ml syringe — 4
- 2 ml syringe — 2
- 1 ml syringe — 4
- Cotton balls — 4
- Spirit/povidone-iodine — 30 ml
- Stethoscope — 1
- Gloves — 2 pairs

Infant feeding tubes
- 5 Fr — 2
- 6 Fr — 2
- Thermometer — 1
- 3 days — 2
- Disposable needles 18 G — 4

Section

C. Equipment for Monitoring

5. Pulse Oximeter
6. Apnea Monitors
7. Blood Pressure Monitors
8. Transcutaneous Bilirubinometer
9. Thermometers
10. Weighing Scales
11. Transcutaneous Blood Gas Monitors
12. Capnograph (End-Tidal Carbon Dioxide Monitors)
13. Ultrasound Machine
14. Cerebral Function Monitor
15. BERA Phone: Automated Auditory Brainstem Evoked Response Audiometry
16. Otoacoustic Emission (OAE) Machine

Chapter 5

Pulse Oximeter

Pulse oximeter is a commonly used device, in neonatal intensive care units for monitoring oxygenation in newborn babies. It is difficult to imagine the era which ended 30 years ago, when the only practical assessment of a patient's oxygenation was the presence or absence of cyanosis. The naked eye evaluation of cyanosis is insensitive to pick hypoxemia even at experienced hands. The introduction of the first blood gas analysers in the late 1950s rapidly revolutionized medical practice. Until recently, measurement of arterial blood oxygen saturation required the direct sampling of arterial blood, which though not difficult was invasive and potentially risky. Furthermore arterial blood gas sampling provides only intermittent monitoring and remains relatively expensive. Fortunately, a major advancement in this field was the development of pulse oximetry to determine percent saturation of hemoglobin with oxygen.

Pusle oximetry technology was available in 1930s but was limited in its use, being cumbersome and bulky. It only became widely available in the 1980s with advances in the light emitting diode (LED), microprocessors, optical plethysmography and spectrophotometry. Today, pulse oximetry provides a simple, non-invasive, portable and inexpensive method to continuously monitor oxygen saturation and heart rate with good accuracy.

In some pulse oximeters the LED and measuring photodiode are on the same side and a reflection technology is used to bounce the light waves back to the same side of the device.

Site of attachment
1. Finger tip
2. Handheld
3. Wrist
4. Table top

PHYSICS OF PULSE OXIMETRY

The concept of pulse oximetry is based on the Beer-Lambert law, which states that the concentration of an unknown solute in a solvent can be determined by light absorption, i.e.

$$L (out) = L (in) - (DCa)$$

where,
- L = Intensity of light
- C = Concentration of solute in the solution
- D = Distance in travelled by light in the solution (path length)
- a = Absorption coefficient of solute.

The core physical principles, based on which pulse oximetry works are:
1. The presence of a pulsatile signal generated by the arterial blood flow.
2. The difference in the absorption spectra of oxyhemoglobin (HbO) and reduced hemoglobin (HbH).

As we are interested in whether oxygen is attached to hemoglobin or not, the relevant solutes are oxyhemoglobin (HbO) and reduced hamoglobin (HbH). The difference in peak absorption characteristics of these two

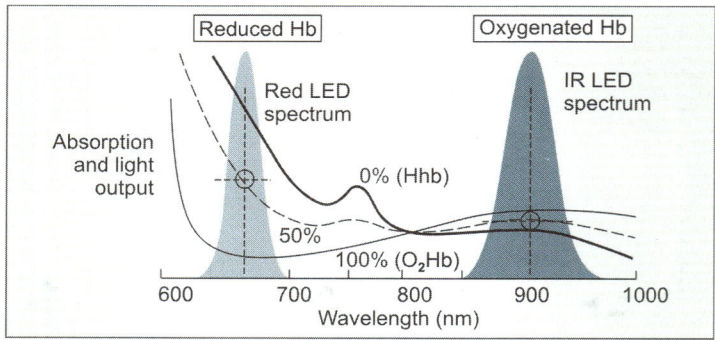

Fig. 5.1: The absorption spectra of oxyhemoglobin and reduced hemoglobin across the red (660 nm) and infrared (940 nm) spectra with middle dashed line depicting spectra of 50% saturated hemoglobin. Reduced hemoglobin absorbs more red than infrared light and oxygenated hemoglobin absorbs more infrared than red

forms of hemoglobin, at wavelengths of 660 nm (red) and 940 nm (infrared), are used in pulse oximetry. *Reduced hemoglobin absorbs more red light than infrared light and oxygenated hemoglobin absorbs more infrared than red* (Fig. 5.1). Further, only the pulsatile change in light transmission through living tissue is measured to calculate arterial saturation with the understanding that such a change in light transmission would solely be due to change in intervening blood volume. Thus, absorption of light by venous blood, skin pigments, tissue and bone is automatically eliminated from consideration.

PRACTICAL WORKING OF PULSE OXIMETER

Probe of pulse oximeter consists of two diodes which emit equal intensities of red and infrared light in sequence into pulsatile tissue bed. Variable amount of these lights are absorbed by oxygenated and reduced hemoglobin. A photodetector placed on the opposite side senses the ratio of red and infrared light based on which the proportion of oxygenated and reduced hemoglobin is estimated by an in built microprocessor and digitally displayed (Fig. 5.2).

Correlation with PaO_2

The PaO_2 at any given saturation is a function of the "oxyhemoglobin dissociation curve" which is sigmoid in shape. As this curve

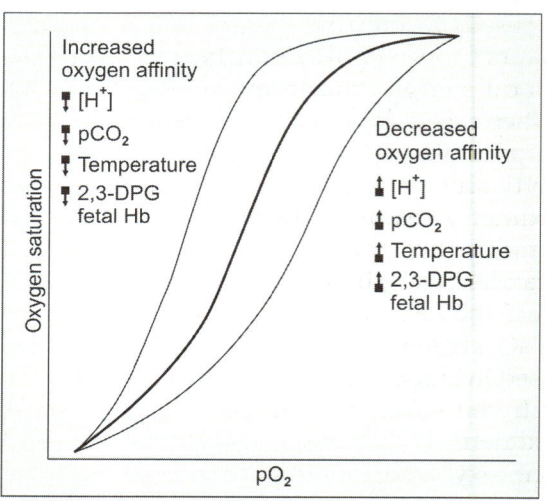

Fig. 5.2: Pulse oximetry theory of operation

reaches flat upper end, further increase in PaO_2 causes a little change in saturation. If pulse oximeter shows high saturation (around 100%), one never knows how high the actual PaO_2 might be. This poor discrimination at the upper end of the curve is accentuated by any shift in the curve itself. A shift to right means less saturation at given PaO_2. Factors which shift the curve to the right are acidosis, high $PaCO_2$, increased temperature, higher concentration of 2, 3-DPG and higher concentration of adult hemoglobin. At <80% sPO_2 precision of pulse oximeter decreases dramatically due to steep curve of O_2 dissociation curve (Fig. 5.3).

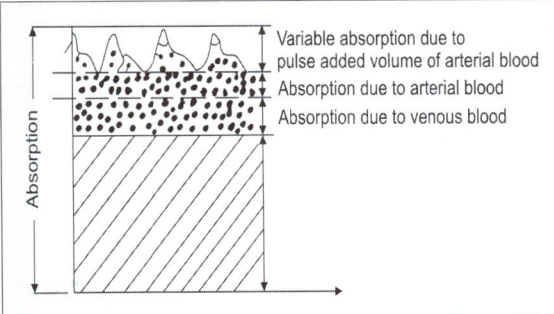

Fig. 5.3: Oxygen dissociation curve

Pulse Oximetry Versus Co-oximetry

Most studies of oximetry performance have focused on comparing single sPO₂ reading to the sPO₂ of simultaneously drawn arterial blood sample measured in a co-oximeter. When this is done accuracy is ± 2%. The more important reliable parameter for monitoring critically ill patients is continuous pulse oximetry reading and trend over period of time rather than single value. The gold standard for estimation of oxygenation, is the determination of partial pressure of oxygen (PaO_2) by co-oximetry, which is the principle used in blood gas analysers. The fundamental difference between pulse oximetry and co-oximetry is that pulse oximetry accounts for only oxyhemoglobin and reduced hemoglobin, whereas in co-oximetry other forms of hemoglobin like methhemoglobin, cyanhemoglobin.

Functional hemoglobin saturation

$$= \left(\frac{\text{Oxyhemoglobin concentration}}{\text{Oxy Hb conc.} + \text{Red Hb conc.}} \right)$$

(Fractional oxyhemoglobin)
↓

$$FO_2Hb = \frac{CO_2Hb}{CtHb} \left(\frac{\text{Oxyhemoglobin conc.}}{\text{Total hemoglobin conc.}} \right)$$

Calibration

Since a ratio rather than absolute value is measured, photo sensors do not need any calibration. However, calibration curves pro-grammed into the software vary from manufacturer to manufacturer and can be different in various pulse oximeters of the same manufacturer. Apart from this there could be some error in the wavelength of the light emitted by the LEDs. For these reasons, same pulse oximeter and probe should be used for all saturation determination in a given patient.

USEFULNESS OF PULSE OXIMETRY

The pulse oximeter has emerged, as simple, non-invasive device, for continuous cardio-pulmonary monitoring in critical care units, emergency departments, etc. The following are its salient uses, with respect to neonatal practice:

1. It provides the fifth "vital parameter", besides temperature, pulse respiration and blood pressure.
2. It is a useful adjunct in the assessment of response to resuscitation.
3. It is an important measurement to aid in titration of oxygen therapy in newborns.
4. It can act as apnea monitor (indicating bradycardia and desaturation).
5. It is a valuable companion during transport of newborns.
6. It may be useful in addition to Allen's test to detect ulnar artery patency.

Situations in which Pulse Oximetry does not Work

The pulse oximetry is often described as a "fair weather friend". It is less accurate in the following situations:

1. Hypovolemic states or low perfusion states.
2. Dyshemoglobinemias—COHb, Meth Hb.
3. Dyes and pigments including nail polish—methylene blue.
4. Optical interference from external light sources (phototherapy unit, fluorescent light, sunlight) and optical shunts.
5. Excess movement artifacts.
6. Excess pigmentation—in dark skinned people it may be less accurate.

7. Pulse oximeter software varies between different manufactures, hence usage of one manufacturer's sensor by another manufacturers gives false results (e.g. Nelcor sensor for Philips monitor).

Fetal hemoglobin and bilirubin most probably do not affect the accuracy of the pulse oximetry.

Pitfalls and Precautions

1. Pulse oximeters are accurate mainly when the oxygen saturation is between 80 and 95%. The accuracy of pulse oximetry is about ±4 to 5% at or above 80% saturation. Accuracy declines below a saturation of 80%.
2. It is mandatory to have a sharp, well defined pulsatile waveform tracing with dicrotic notch, for the saturation and heart rate readings to be accurate.
3. Interference from other light sources gives erroneous readings, which can be avoided by covering the pulse oximeter probe with opaque material like aluminum foil.
4. Movement by the newborn baby may lead to a disrupted signal and artefacts.
5. Avoid compromising blood flow to the limb as well as pressure necrosis of the site, to which the probe is attached to prevent a false low readings as well as injury to patient.
6. If probe does not fit properly, the light can be shunted from the LEDs directly to photodetector affecting the accuracy of the measurement.
7. Pulse oximeter is not reliable in conditions of severe hypotension or severe hypothermia (in such conditions an ear probe may be more reliable than a finger probe).
8. Pulse oximetry does not take clinical impact of anemia into account; hence it is less accurate in severe anemia.
9. Currently available pulse oximeters are unable to distinguish different types of hemoglobins. Hence, in the presence of COHb (carboxyhemoglobin) and Meth Hb (methemoglobin), the saturation readings may be falsely and significantly elevated, thus masking the presence of hypoxemia.
10. Always remember, pulse oximetry reflects state of oxygenation. It has no value in estimation of adequacy of ventilation.
11. *Lag monitor phenomenon*—often a fall in partial pressure of oxygen precedes the fall in oxygen saturations, by several minutes and there may be a delay in picking hypoxemia episode.
12. *Response delay*—although described as device for continuous real time monitoring of oxygen saturations, there is often a delay of 5–20 seconds due to signal averaging. Hence, actual drop in saturations precedes the displayed drop.

SIGNAL EXTRACTION TECHNOLOGY (MASIMO)

Masimo signal extraction technology (SET), patented by Masimo corporation, USA, enables accuracy of sPO_2 measurement during low perfusion states, movement and noise artifacts. While conventional pulse oximetry employs one, or two algorhithms to attempt to measure patient arterial saturation, Masimo SET employs five algorithms using adaptive filters, working in parallel to ensure accurate measurement in difficult situations like motion or low perfusion (Fig. 5.2). Further there are now next generation pulse oximeters by Masimo—the rainbow SET pulse co-oximetry, which uses seven algorithms, enabling this device to measure non-invasively, total hemoglobin (Sp Hb), respiratory rate (RRa), pleth variability index (PVI), oxygen content, and levels of carboxyhemoglobin (SpCO) and methemoglobin (Sp Met).

Points to Remember

i. Desired O_2 saturation will vary according to the infant's condition. Physician should specify the desired range which is as follows:
   - Premature (1–2 week)—90–93%
   - Older neonate, especially with BPD—90–95%

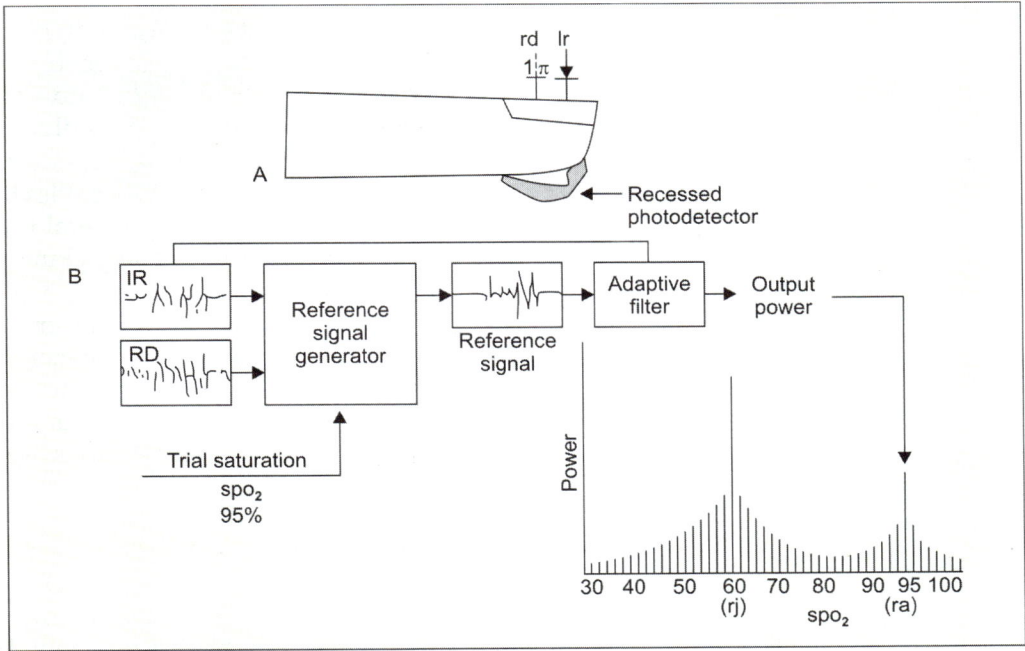

Fig. 5.4: Masimo signal extraction technology (a patented technology of Masimo corporation USA), using four extra algorithms in addition to the red/infrared ratio used in conventional pulse oximeter algorithms and adaptive noise filter, which enables accurate measurement in low perfusion states and negates motion artifacts or excess ambient light. (*Courtesy:* Masimo corporation, http://www.masimo.co.uk/whymasimo/difference.htm)

ii. Alarm limits are kept 2% higher and lower than the desired saturation range.
iii. Inaccurate readings may be due to:
- Poor tissue perfusion
- Cool periphery (cold stress/hypothermia)
- Exposure of probe to light sources
- Excessive movement of limb
- Electrical interference from other equipment
- While taking BP from the same limb/any obstruction to blood in that limb, e.g. a splint tied tightly for IV access

iv. Oxygen saturation monitors are unreliable in detecting hyperoxia at high saturation values.
v. The error associated with saturation monitor reading is 2% in the range of 95–100%. Therefore, a saturation reading of 96% may be as low as 94% or as high as 98%.

Complications of Pulse Oximetry

These are rare and include finger burns and pressure necrosis due to prolonged contact with probe.

Pre-requisites of a Good Pulse Oximeter

The main consideration, when buying a pulse oximeter is the cost incurred. A good device should have the following features:

1. The display should indicate a pulse-wave form, and heart rate with in-built alarm limits for the heart rate and saturation.
2. There can be additional features like adjustments in pulse volumes and alarm volumes.
3. The set should be tropicalized for use in India.
4. It should be small enough to be portable, with an in-built battery pack, which should be able to provide power for at least 4 to 6 hours. This battery should have a short recharging time.

5. The probe should be flexible (flexiprobe) and usually is cheaper than a finger probe. There should be at least one or two spare probes (Figs 5.5 and 5.6).
6. A good backup/maintenance service is as mandatory.

Depending on financial constraints, one can make use of different combinations of features to obtain a cheap but useful instrument. Obviously, long duration trends, storage

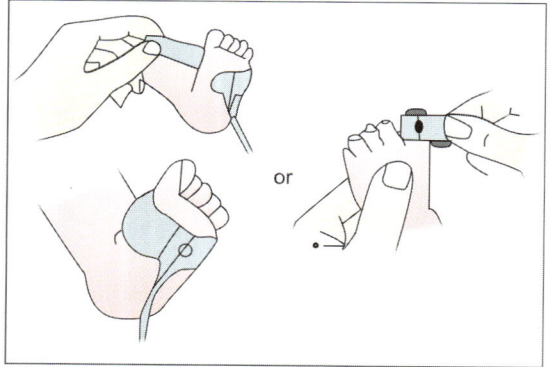

Fig. 5.5: Application of oxygen saturation probe on feet

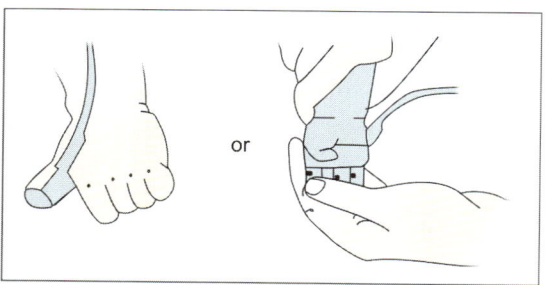

Fig. 5.6. Application of oxygen saturation probe on hand

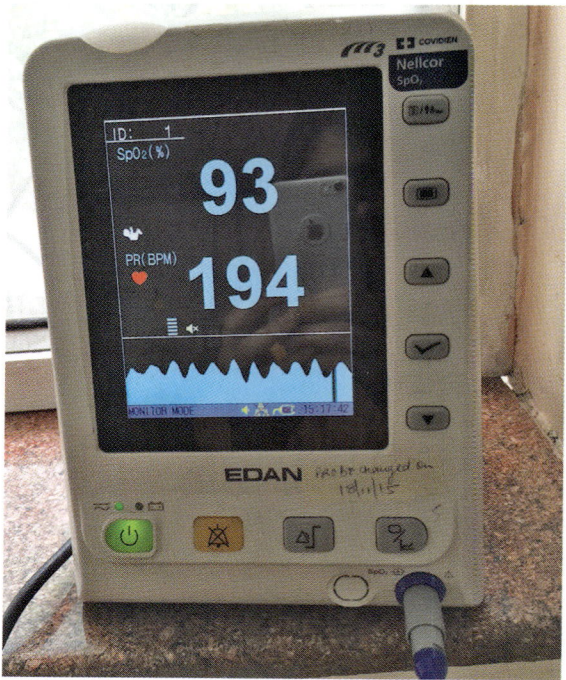

Fig. 5.7: Pulse oximeter

facilities of data and printer options may be useful though rather expensive (Fig. 5.7).

Finally it is important to remember that pulse oximetry monitoring serves as a useful parameter, complementing clinical examination, rather than replacing it.

Various portable low cost battery operated finger pulse oximetry from various companies (₹ 1000–2000) are available in Indian market. They cannot be reliably used for newborns. However, they may be useful during transport (Table 5.1).

TABLE 5.1: Common brands available in market

S.No.	Make	Dealers	Unit cost (₹)
1.	Pacetech model 520, Series 300	Medex	50,000–90,000
2.	Novamatrix model 511, 515 C Oxypleth	Rustagi Surgicals	40,000–60,000 70,000
3.	Nellcor N-180, 185	Instromedix	57,000–80,000
4.	Simed	Oticare	70,000
5.	Oxypal Nihon Khoden	Shibumi Medical System	1,35,000

(Contd...)

TABLE 5.1: Common brands available in market (*Contd...*)

S.No.	Make	Dealers	Unit cost (₹)
6.	Miniox	Mediserve	50,000
7.	Criticare 503, series 507	Criticare India Ltd.	80,000
8.	Datex	AVL Biomedical	75,000
9.	Ohmeda Biox-3700	Datex-Ohmeda	1,75,000
10.	Invivo	Instrument and machine	70,000
11.	Minolta	Draeger consolidated products	70,000
12.	Pulseox	Methodex	80,000
13.	Masimo	Innovative Intex Pvt. Ltd.	1,00,000
14.	Mediaid	Cirticare, UK Allied Health, Gurgaon	80,000
15.	Comet-P	M/S cardio care/Larsen and Turbo	75,000
16.	Oxywave	Schiller	80,000
17.	PM 100N	Covidien, Nellcor	70,000
18.	BCI Autocorr	DOT med India, Smith Medical	60,000
19.	Mediana sPO$_2$ monitor	Digitex Medical, Mediana China	60,000
20.	Schiller Truscope Mini	Schiller India, Schiller Germany	50,000

Frequently Asked Questions (FAQs)

Q. 1. At what site should the probe be placed for pulse oximetry?
The probe can be positioned on the fingers, toes, hand, foot, or wrist of the neonate. Other sites will depend on the infant's size. Newer probes allow for forehead placement. The placement of ear lobe probes is particularly useful in hypoperfusion states.

Q. 2. Can ambient light interfere with pulse oximetry readings?
Yes, ambient light containing the red spectrum may interfere with accurate readings from the oxygen saturation monitor. Light from heat lamps and phototherapy lights has been reported to skew the readings. The high intensity of light emitted from these sources masks the small changes in light transmission from the probe. The remedy is to shield the probe from the ambient light by black paper (e.g. carbon paper) or black polythene or aluminum foil.

Q. 3. How can I determine the accuracy of the pulse oximeter?
Most pulse oximeters have a visual representation of the pulse intensity as well as a digital display of the pulse. The pulse display should be within three beats per minute of the display on the cardiac monitor. The bar pulse display or pulse waveform must cover half of the total display for an accurate reading. Differences greater than this will not reflect accurate oxygen saturation values because the probe is not detecting the arterial pulsations adequately or accurately. Some newer monitors have integrated the ECG complex with the oxygen saturation probe.

Q. 4. Which infants can be monitored using the oxygen saturation monitor?
The oxygen saturation monitor is reliable, practical, and accurate for use in infants with a wide range of birth weights, gestational age, postnatal ages and heart rates.

Q. 5. What is purpose of pulse oximetry?
It helps in
- Noninvasive arterial oxygen saturation monitor.
- Pulse rate monitoring.
- Trending of saturation and pulse rate.

Q. 6. Common indications for pulse oximetry.

1. To measure oxygenation in infants suffering from hypoxia, apnea, cardiorespiratory disease, bronchopulmonary dysplasia.
2. To monitor response to therapy during resuscitation:
   - Resuscitation
   - Monitoring effectiveness of bag and mask ventilation or during placement of endotracheal tube.
3. Monitoring side-effects of therapy:
   - Suctioning
   - Laryngoscopy
4. For extreme LBW babies <1000 g, to restrict and regulate oxygen delivery.
5. During neonatal anesthesia to regulate FiO_2.

Q. 7. Are there any complications from using the oxygen saturation monitor?

In the newborn population, there are no known complications from oxygen saturation monitoring when the neonatal probes are used as indicated.

Q. 8. How do I clean/sterilize the probe after use on one baby and before being used on another?

Cleanse the probe with alcohol, let it dry before using on another baby.

Q. 9. What are the comparative advantages and disadvantages of the transcutaneous oxygen monitor and the oxygen saturation monitor?

A comparison of the two monitoring systems is outlined in the table given below.

Q. 10. What are the responsibilities of staff for using pulse oximetry?

1. Calibrate with arterial blood gases if applicable for model and brand.
2. Select the appropriate sized probe, and locate a position for monitoring.
3. Place the probe such that the light source and photodetector are opposite each other.
4. Set monitor alarms in accordance with the unit policy.
5. Document monitor readings and FiO_2 every hour and with each blood sample drawn for gas analysis.
6. Change probe site to avoid skin breakdown.

Q. 11. What are the common brands available in the Indian market?

You may choose anyone of the following or other brands available. A few may be very cheap, like a handheld pulse oximeter which runs on battery alone. It does not have display of pulse waveform/bar. The disposable probes will be cheap, but recurring cost will be quite high. Reusable flex probe (life 3–6 months costs only one-third) of finger probe (life 1–2 years; cost ₹ 6,000 to 8,000/-).

Feature	Transcutaneous oxygen tension	Pulse oximetry
Calibration	8–10 mins every 4 hours	None
Warm-up-time	Approximately 15 minutes after probe application for the skin to reach 43°–44°C and capillary bed to "arterialize"	None (displays oxygen saturation level instantly once a pulse is located)
Lag time complications	30–40 seconds. Thermal injuries resulting in first and second degree burns due to heat generated from probe	None (instantaneous); compromised skin integrity under the probe (may go unnoticed because the probe does not need repositioning at set intervals)
Artifacts (factors causing inaccurate readings)	Membrane wrinkles, air between the membranes and skin, pressure on the probe	Movement of extremity with probe; inflated blood cuff proximal to probe; light with red spectrum reaching an unshielded probe

Note: There is no doubt that pulse oximetry is remarkably more practical and useful option.

Q. 12. What are steps involved in setting up a pulse oximeter?

Procedure	Rationale
1. Assemble all necessary equipment	
2. If saturation monitor probe is reusable, cleanse probe with alcohol, let it dry	Decreases risk of nosocomial infection and facilities transmission of light
3. Turn monitor on	
4. Apply probe to a site that is well perfused	Pulse oximetry requires adequate perfusion of tissues and pulsating vessel to transmit red light
5. Ensure both sides of probe are directly opposite each other	One side emits light the other side has photodiode that picks up the transmitted light
6. Secure probe in place. Avoid edematous, bruised sites and excessive pressure	
7. Set high and low alarm limits for saturation and heart rate (2% above and below desired limits)	Warns of excessive or low oxygen saturation and heart rate
8. Set pulse and alarm volumes	
9. Check for correlation of depicted heart rate on monitor and the actual heart rate by auscultation	Saturation readings are likely to be correct if heart rates on cardiorespiratory and oxygen saturation monitors correlate
10. Record heart rate, respiratory rate, color, oxygen saturation and FiO_2 hourly	
11. Observe and change site at least once per shift	Prevents skin breakdown

Q. 13. What can go wrong with the pulse oximeter setting?

Problems	Remarks
1. The oximeter continues to "search" but cannot find a pulse, or there is a pulse displayed, but no % sat displayed	Try readjusting the sensor or applying it to a new site which has better perfusion. Slight readjustments in the position of the light emitter and/or light detector can make the difference between a non-functional and a functioning sensor. Be sure too, that the "windows" are clean. Tiny bits of debris can interfere with the sensor. The sensor may not be plugged in securely to the monitor, or the sensor may be damaged.
2. The heart rate and % sat readings fluctuate rapidly	This is usually motion artifact caused by an active baby, although some of the fluctuations may be real
3. The % sat readings are misinterpreted to be the PaO_2, as determined from an arterial blood sample	Remember that PaO_2 also measures the oxygen dissolved in plasma, which is reported as "mm Hg". An oximeter measures oxygen bound to hemoglobin, which is reported as "% sat." Although the two values should correlate (*see* the basic unit) the exact numbers will not be the same.
4. The oximeter reading seems to be inaccurate	Check to be sure that the pulse oximeter pulse rate is the same as the heart rate as determined by a cardiac monitor. If they are different the sensor may need to be adjusted. Also, check to be sure the sensor is shielded from bright light. Finally use your clinical acumen and evaluate the baby rather than just concentrating on the machine

Q. 14. What makes Masimo SET different from conventional pulse oximetry?

Conventional pulse oximetry assumes that it is the arterial blood which moves in a pulsatile manner but in actual is the venous blood too also moves which is under read as the conventional pulse oximetry fails to distinguish venous from arterial blood. Masimo SET identifies the venous blood signals, isolates it, and using adaptive filters , cancels the noise and extracts the arterial signal. This technology will work during patient motion artifact , low persfusion states with low signal output and during intense ambient light exposures. So this technology filters out the desired signal while discarding the unwanted noise signals, thereby making the saturation estimate more accurate in situations where conventional pulse oximeters fails.

Q. 15. What is Masimo rainbow SET technology?

This is the latest addition to the newer generation signal extraction technology pulse oximeters, where adaptive filters along with seven different algorithms, are used to non-invasively measure blood constituents and fluid status using pulse oximeters. It claims to measure, total hemoglobin (Sp Hb), respiratory rate (RRa), pleth variability index (PVI), oxygen content, and levels of carboxyhemoglobin (SpCO) and methemoglobin (Sp Met).

Chapter 6

Apnea Monitors

Apnea is defined as cessation of breathing for more than 20 seconds with or without bradycardia and/or cyanosis. Cessation of respiration for 10 seconds with bradycardia or cyanosis may also be considered as apnea. Apnea may further be classified as central, obstructive or mixed apnea. Around 20–25% of preterm babies weighing less than 1500 gm may develop apneic episodes. Prompt recognition is important to prevent hypoxia in the neonate. Treatment is in the form of physical stimulation, drug therapy and occasionally ventilatory support. Since it is not possible to observe these babies constantly, monitoring equipment should be used to detect apneic episodes in all babies less than 32 weeks and less than 1500 gm. These monitors are called *apnea monitors*.

Apnea monitors may be invasive or non-invasive. Invasive or direct monitors would actually measure airflow in the airway through an endotracheal tube. Non-invasive monitors either detect chest wall movement or measure air flow through oral/nasal cannula. Non-invasive monitors are popular for obvious reasons.

Conventional apnea monitors which detect chest wall movement will be discussed. These can pick up central apnea but fail to pick up obstructive apnea. Therefore a combination of both chest wall movements and airflow detectors would be ideal. The most cost-effective and common monitors are those which are based on thoracic impedance.

Basic Principle

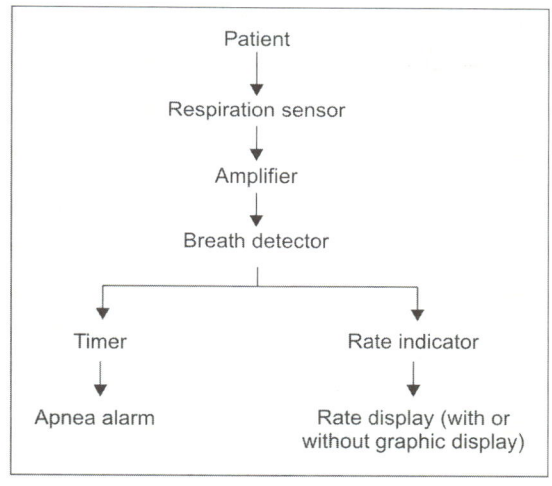

A respiratory sensor indicates whether the baby is breathing or not. If breathing, the rate is measured. A continuous display of the respiratory rate can be obtained. Alarms are activated if the rate goes above or below a preset value. The sensor and breath detector is also attached to a timer and the alarm system is activated if the sensor does not detect any respiratory movement for a preset period of time, i.e. usually 20 seconds. Thus, the alarm indicates whether the baby has a high or low respiratory rate and if a baby does not breathe for a preset period of time, usually 20 seconds. One could set the timer to act as a visual display. The intensity of sound can be altered. Some monitors also have a heart rate display and alarms can be set for high and low heart rates.

What are some features of apnea monitors?

- *Battery back-up*: Most apnea monitors can be plugged in, or powered by a battery. The apnea monitor may have a battery backup. If the power goes out, the battery will give power to the device to help it keep working.
- *Environmental seal*: This special covering protects the device from cold, heat, wetness, and insects.
- *Memory recorder*: This is also called the event recorder. This feature keeps a record of the times, when baby has apnea, or the times when babies oxygen blood level has gone too low.
- *Modem*: This lets you send information from the apnea monitor to your caregiver using a telephone line.
- *Monitoring*: An apnea monitor will alert when it is not working right. It may show a blinking light, or make a sound. The monitor will sound an alarm as mentioned previously.

TYPE OF APNEA MONITORS

There are different types of apnea monitors depending on how they pick up chest wall movements.
1. Movement sensors
2. Thoracic impedance monitors
3. Respiratory inductive plethysmography
4. Magnetometer
5. Passive apnea monitor.

1. Movement Sensors

i. *Apnea mattress*: Ripple type mattress where air is displaced from one section to another when the baby breathes. Airflow is detected over a thermistor which gives a feedback regarding presence or absence of respiration. This is suitable for tiny babies. The mattress should be properly inflated.
ii. *Mattress with sensory pad*: Baby lies on a mattress with a sensory pad which senses pressure changes from respiratory movements using a plethysmographic sensor. Non-respiratory movements like cardiac pulsations could also be detected which is source of error.
iii. *Pressure sensitive capsule*: The capsule is attached to the skin which gives feedback regarding presence or absence of breathing. *In general, monitors based on movement sensors will not detect obstructive apnea and may not distinguish body movements from breathing.*

2. Thoracic Impedance Monitors

Transthoracic impedance measurement is the most commonly and widely used method for monitoring respiration. The technique involves the passage of a low level, high frequency signal through the patient's chest. The signal is passed by the same electrode used for cardiac monitoring. In most cases the RA (right arm-white) and the LA (left arm—black) are used as the path for the high frequency signal (Fig. 6.1). As the density of

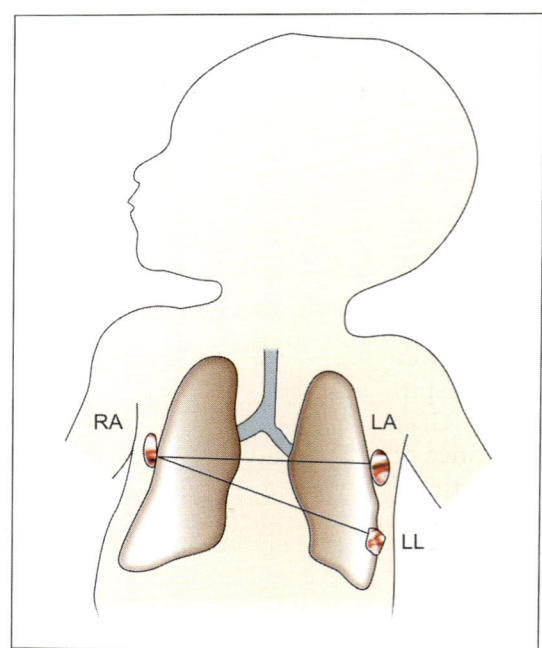

Fig. 6.1: Transthoracic impedance pneumograph. Path of the high-frequency signal through the patient's chest. *Notice*: Most monitors use right arm → left arm; others may use right arm → left leg

the chest cavity changes due to lung inflation, the impedance of chest cavity also changes (impedance = the electrical resistance to high frequency signal). The high frequency signal is transmitted through the chest and undergoes an amplitude change proportional to the impedance changes of the thorax. This impedance change, as seen by the modulation of the high frequency signal, is detected and quantified by the monitor and counted as respiratory activity per minute or breaths per minute. The monitor will set an impedance threshold limit for valid respiratory activity because cardiac pumping (pulmonary blood flow) action will also cause thoracic impedance changes (usually much smaller than respiration changes). The monitor will separate the ECG signal from the high frequency carrier signal used for detection of respiration.

Electrodes are placed directly on the infant's chest or inside an adjustable belt that is secured around the chest. A small electric current is passed between the electrodes. The impedance changes with breathing because of air-fluid shifts. This change in transthoracic impedance detects presence or absence of respiration. Heart rate can also be detected using these monitors. Like the movement sensors this also cannot detect obstructive apnea. Cardiac activity fluctuation in impedance may be mistaken for respiration.

3. Respiratory Inductive Plethysmography

This uses abdominal and thoracic movements during respiration. Abdominal thoracic bands or the Graseby capsule are used, the inductance of which changes with breathing. This method is more sensitive in picking up obstructive apnea and is less affected by cardiac artifacts.

4. Magnetometer

A magnetometer can be taped on the anterior part of the chest or abdominal wall. A magnetic field sensor such as an electric coil is then placed under the infant. Electrical signals are produced due to the chest or abdominal wall movement. This can be detected by the monitor. This type of monitor is more sensitive, but more expensive and hence not widely used.

5. Passive Apnea Monitor

The instrument monitors, the acoustic and electromechanical signals of the patient and calculates an energy spectrum periodogram or histogram using time series analysis techniques. The patient lies down on a large piezoelectric film (few microns thick) that has the capability of measuring signals from very high to very low frequencies. The heart and respiration rates as well as obstructive apnea can be observed, detected and measured from the spectral peaks in the resulting energy spectrum. A micro-computing machine provides calculations to determine the energy spectrum and provides for discrimination between noise and a true apnea episode. An alarm calls for assistance in the event of an apnea, including obstructive apnea, or "acute life-threatening event (ALTE)".

How to use an apnea monitor?

1. The mattress is easy to use. The baby is made to lie on the mattress and the monitor switched on.
2. In the impedance monitor, high and low respiratory rates are set usually at 60/min and 40/min. Apnea alarm is set at 20 seconds. The loudness of the alarm is adjusted. The electrodes are placed at the side of the chest wall using disposable neonatal electrode pads and the monitor is switched on. Care should be taken to ensure that leads are fixed properly. Most monitors have a loose lead alarm and this can be rectified easily. Electrode pads should necessarily be disposable and small in size.

How does an apnea monitor look like?

An apnea monitor often looks like a box with knobs, wires, TV-like screens, and a power cord. Some apnea monitors are small and light enough to be carried easily from place to place. Commonly, electrodes are placed on each side of chest. They may stick on to the skin, or be held on with a light weight belt. There may also be other attachments that measure air flow through the nose or mouth. The electrodes have wires coming from them which are attached to the device.

Advantages

1. Easy to operate.
2. Detect apnea.
3. Time limits can be set.
4. Respiratory rates can be obtained.
5. High and low respiratory rates can be set (alarm limits).

Disadvantages

1. May not detect obstructive apnea, can be overcome by using a passive apnea monitor.
2. May pick up cardiac artifacts.
3. Movement sensor type may not distinguish cardiac pulsations from respiratory movements.

Recent Trends

- Combination of conventional monitors with airflow sensors to detect both kinds of apnea.
- Apnea monitors attached to a "stimulator"—so that when apnea is detected not only does the alarm get activated, but the baby also gets stimulated.
- Combination monitors may be better than isolated apnea monitors, i.e. monitors which detect apnea, heart rate, saturation, etc.

Frequently Asked Questions (FAQs)

Q. 1. If one could choose, which apnea monitor should one purchase?

Apnea monitor systems with simultaneous heart rate recording along with respiration are ideal. In the situation of obstructive apnea, bradycardia will alert the staff, while an apnea monitor based on chest wall movement detection is likely to miss this event.

Q. 2. Which babies need to be on an apnea monitor?

1. All babies less than 32 weeks should be monitored for at least the first week of life. Any sick baby admitted in the intensive care unit ideally should also be on an apnea monitor.
2. Babies with bronchopulmonary dysplasia and pneumonia—as these infants are more prone to develop apnea.
3. *Sudden infant death syndrome (SIDS)*: Babies with history of previous sibling death due to SIDS can be monitored at home.

Q. 3. Which are the cheapest and reliable apnea monitors?

Cheaper apnea monitors work on the principle of movement. An apnea monitor using Graseby capsule attached to the abdominal wall detects each respiratory movement (audible and visual display) and gives alarm if there is no movement (apnea delay set time). Similar is the functioning of an apnea monitor mattress kept below the baby which works by sensing pressure changes from respiratory movements using a plethysmograhic sensor (Table 6.1).

Q. 4. What are the reasons for false signals in transthoracic impedance pneumography?

False positive signals in absence of effective ventilation may occur due to:

i. Chest wall movement with airway obstruction.
ii. Nonrespiratory muscular action giving motion artifacts—stretching, seizures, etc. A false alarm may be due to improperly set sensitivity; not detecting respiratory activity; electrodes not correctly placed and loose electrodes.

TABLE 6.1: Apnea monitors available in the market

S.No.	Make	Principals	Dealers	Unit cost (₹)
1.	Mattress type	Lektromedik Rustagi Surgicals	Lectromedik	₹ 15,000 to 1,00,000
2.	Apnea monitor (RR alone, or with HR)	Meditrin Phoenix	Meditrin Phoenix	₹ 12,000 to 15,000
3.	Graseby capsule	SIMS Graseby	Medisphere	₹ 45,000
4.	Neo-Trak 500 (only apnea) Neo-Trak 502 (with HR/RR)	Carometrics	Micronic device	₹ 70,000 to 1,00,000
5.	Neoguard 280	GE	Wipro GE Healthcare	₹ 50,000
6.	Health dyne apnea monitor 900	Health dyne	Elder Health care	₹ 1,00,000
7.	System V/VI	GE	Wipro GE Healthcare	₹ 1,50,000 to 3,00,000

See chapter on blood pressure monitors

Q. 5. Can one use pulse oximeter as substitute for an apnea monitor?

Yes. A pulse oximeter in which the alarm gets activated once the baby has bradycardia (HR <100/min) or desaturation (saturation <88–90%) may be a good and reliable alternative to an apnea monitor.

Q. 6. Why is a home apnea monitor used?

A home monitor can be used to monitor babies who still have minor problems with breathing and heart rate. However, studies have not been able to show any benefit for babies on home monitors, so the monitors are not commonly used anymore. There may be special situations where one may be necessary, but most babies do not use home apnea monitors. Few indications include infants who have persistent apnea or severe reflux, a family history of SIDS, or who need home oxygen or on respiratory support.

Chapter 7

Blood Pressure Monitors

Blood pressure, as the name implies, is the force exerted by the blood against any unit of the vessel wall, at the moment when it is measured. This is conventionally expressed in millimeter of mercury (mm Hg). The numerical value of the systemic arterial pressure is determined by the extent of filling, that is, the fullness of the arterial side of the systemic vessels and the intrinsic resistance of these vessels to this volume of blood. This value is constantly changing owing to the pulsatile nature of filling and run-off related to the cardiac cycle. Blood pressure is thus, a product of peripheral vascular resistance and cardiac output.

Systolic blood pressure (SBP) is the pressure at the height of the arterial pulse. It coincides with the left ventricular systole. The systolic pressure depends on the stroke volume, cardiac contractility and arterial dispensability. *Diastolic blood pressure (DBP)* is the lowest point of arterial pulse and coincides with the left ventricular diastole. The diastolic pressure depends on the level of systolic pressure, peripheral resistance, heart rate and arterial elastic recoil. **Mean blood pressure (MBP)** is determined by the integration of pulse contour during the interval of the cardiac cycle. It is roughly the diastolic pressure plus one-third of the pulse pressure, where pulse pressure is the difference between systolic and diastolic pressures.

NORMAL BLOOD PRESSURE

This is defined physiologically as the blood pressure range which ensures optimal supply of oxygen and nutrient to tissues without causing damage. Blood pressure is monitored in all neonates who are at any risk of hypotension or hypertension. Normal blood pressure varies by gestational age, body weight, cuff size and state of alertness. Normal values have been developed by body weight and postnatal age. It is important to display the norms of blood pressure in the NICU so that values obtained in a given baby could be readily compared and assessed as normal, high or low enabling the staff to set alarm limits as well as take appropriate clinical action. Generally, a neonatal diastolic pressure lower than 25 mm Hg and a mean pressure less than 30 mm Hg should cause concern. Similarly, a DBP >50 mm Hg in preterm and >60 mm Hg in term should be considered beyond normal range. The aim of monitoring is to maintain the mean blood pressure within normal range. For practical purposes these ranges are given in Table 7.1.

TABLE 7.1: Mean blood pressure based on birth weight

Birth/weight	Mean blood pressure
<1 kg	30–35 mm Hg
1–2 kg	35–40 mm Hg
2–3 kg	40–45 mm Hg
>3 kg	45–50 mm Hg

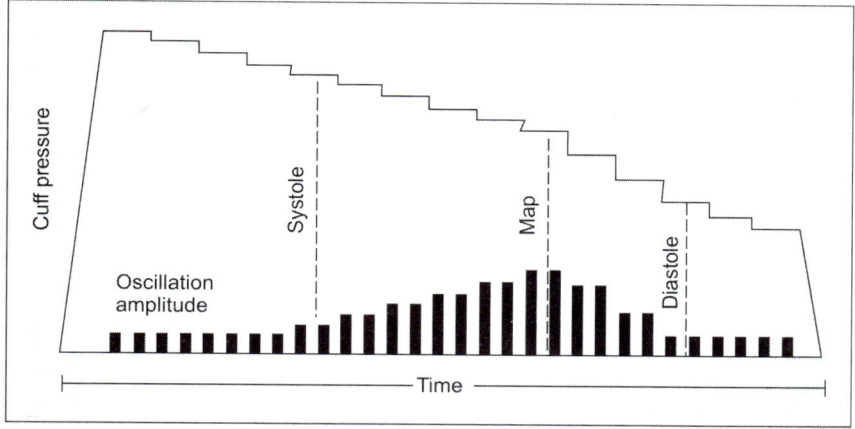

Fig. 7.1: Determination sequence for oscillometric measurement

There are two methods for measuring blood pressure: Non-invasive and invasive (arterial cannulation pressure transduction). A direct measurement from an indwelling arterial catheter is the gold standard for measuring blood pressure in newborn. However, correlation between direct and indirect method is generally good. Indirect determinations are generally higher than direct methods, often by 3 to 5 mm Hg. These differ in terms of physical processes, they monitor and the level of invasiveness (Fig. 7.1).

Noninvasive Methods

All indirect measurements rely on the use of inflatable cuff or arterial occlusion and digital or aural assessment of the return of arterial flow. Automatic devices use oscillometric sensors, piezoelectric microphones or Doppler probes to detect return of flow signals.

For accuracy in indirect measurement
1. Place the infant supine, with the limb fully extended, level with the heart.
2. Measure the circumference and select appropriate size cuff for the limb (Table 7.2). Usually 2.5 cm width for preterms and 4 cm width for terms will suffice.
3. Cuff bladder should fully encircle the arm; if it encircles part of the circumference, then the bladder should lie over the artery.
4. Cuff should be applied firmly and not loosely to the arm.
5. Inflation can be done quickly to a pressure of 15 mm Hg above the point at which the distal pulse disappears.
6. Deflation should be slow, 2–3 mm of Hg per heartbeat.

Indications
1. For at risk and unstable babies admitted in nursery.
2. During management of serious illnesses like sepsis, birth asphyxia or hypothermia.
3. During treatment with drugs causing hypotension (tolazoline, pancuronium, morphine) or hypertension (steroids).
4. Babies on inotropes, e.g. dopamine, dobutamine and epinephrine.

Technique
1. Place the infant supine, with limb fully extended, level with the heart.
2. Measure the circumference and select the appropriate size cuff for the limb (Table 7.2).

TABLE 7.2: Neonatal cuff size

Cuff no. (size)	Limb circumference
# 1	3–6 cm
# 2	4–8 cm
# 3	6–11 cm

3. Apply the cuff snugly to the bare limb, above elbow or knee joint.
4. Place the stethoscope or Doppler over the brachial artery or popliteal artery.
5. Inflate the cuff rapidly, to a pressure 15 mm Hg above the point at which the distal pulse disappears.
6. Deflate the cuff slowly.

Complications and Problems

1. *Inaccurate measurements*
   a. Defective pressure measurement system
   b. Air leaks at connections or tubing
   c. Inappropriate cuff size: A narrow cuff gives a high reading; a wide cuff gives a low reading.
   d. Cuff applied loosely leads to an erroneously high reading.
   e. Rapid deflation of the cuff will miss 'the return of flow' point.
   f. Active or agitated patient.
2. Nosocomial infection following use of the same cuff on multiple patients.
3. Pressure may not be detectable in low perfusion state or shock.
4. Prolonged repeated cycling has been associated with ischemia, purpura, and/or neuropathy due to compression of vessels and nerves.
5. Pressure not detectable or inaccurate in neonates experiencing convulsions or tremors.

Methods

i. ***Palpatory***: Palpate for radial pulsation while deflating the cuff. The point of palpable pulsation depicts systolic BP. It is difficult to palpate the pulse in a struggling neonate. Moreover, this does not give any idea of the diastolic BP.

ii. ***Flush***: Place a suitable cuff around the forearm or the calf. Raise and squeeze the limb, while quickly inflating the cuff to about 110 mm Hg. Now lower the limb and place it by the side of the body. Deflate gradually in steps of 5 mm Hg each and wait for 5 seconds at each step. The point at which flushing of the limb appears is the systolic BP.

iii. ***Auscultatory method***: This is based on the detection of Korotkoff sounds during cuff deflation by auscultation. The SBP corresponds to Phase I Korotkoff sound and DBP corresponds to Phase IV Korotkoff sound. Auscultation is difficult in neonates; hence this method is not of much use in them.

iv. ***Doppler ultrasound techniques***: These automated non-invasive BP (ANIBP) devices use ultrasonic technology in the detection of arterial activity during cuff deflation. One type uses the Doppler principle to determine the blood flow distal to the cuff and another senses the motion of arterial wall. An ultrasonic transducer is placed under the cuff which detects the arterial wall movement, the frequency of the reflected sound will be shifted in relation to the incident sound (Fig. 7.2).

By electronically sensing the shift in frequency of the reflected ultrasound and converting it to an audible signal, it is

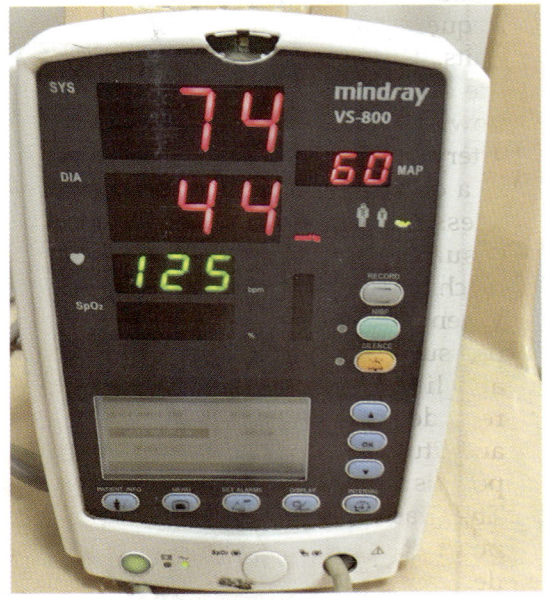

Fig. 7.2: NIBP monitor

possible to indicate the arterial wall movements. Two movements occur: One, during arterial expansion when blood moves through it as the cuff is being deflated, and the second, as the artery collapses when the arterial pressure falls below the cuff pressure. These two shifts are then used by ultrasonic kineto-arteriography to detect the SBP and the DBP.

Advantage is that ultrasound sensors are unaffected by ambient noise (crying of the baby, etc.); hence, they can be used in a noisy environment and only the sound level of the instrument has to be increased. But the major disadvantage is that the ultrasound system is sensitive not only to the arterial wall motion but also to any other structure and the transducer itself. Thus, for reliable results, it is necessary that the subjects be relatively still which is somewhat difficult in neonates.

v. *Oscillometric technique*: Oscillometric technique is highly accurate and is used most frequently in the contemporary non-invasive BP monitors. Dinamap (Critikon Inc., Florida) is an automatic oscillometric blood pressure monitoring device that is frequently used in newborn intensive care units. This is based on the principle that the arterial wall oscillates when blood flows in a pulsatile fashion through the artery. These oscillations are transmitted to a cuff placed around the limb. As the pressure within the cuff is reduced, (usually done automatically by the machine) the pattern of oscillation changes. When arterial pressure is just above cuff pressure there is a rapid increase in the amplitude of oscillation, and this is recorded as systolic pressure. When the amplitude of oscillations is maximum that point is recorded as mean arterial pressure. The diastolic pressure is identified at a point at which amplitude of oscillations decrease suddenly. This apparatus consists of a single cuff connected to a display and control box. In the box, a tube is connected, to a pump, which inflates, and a bleed valve that deflates the cuff and another tube connected to a pressure transducer. The output from the pressure transducer is connected to an AC and DC coupled amplifier. The output of the DC amplifier produces voltage that is a function of the pressure in the cuff. The AC coupled amplifier is used to detect oscillations which are used for calculation of systolic, mean and diastolic BP from the above sequence.

Precautions
1. Carefully select the appropriate cuff size because incorrect size can significantly alter the blood pressure value obtained.
2. The patient must be still during measurements, inaccurate measurements occur due to motion.
3. Cuff should be ideally single patient use to avoid cross infection.
4. Caution should be exercised in interpreting values when used in preterm VLBW infants in a hypotensive state. It tends to overestimate pressure in hypotensive VLBW infants.

vi. *Finger plethysmography*: This uses a small cuff incorporated with a light source and a sensor that encircles a finger. The volume of the finger is maintained constant by continuously adjusting the volume of the cuff and the pressure in it. The volume of the finger is measured photometrically. Under these conditions, the pressure in the cuff is a function of the arterial BP in the finger. This is presently marketed by Ohmeda Finapres for use in adults. It holds great promise in future as this is the only non-invasive system that provides continuous, real time BP monitoring.

vii. *Invasive monitoring*: This requires the use of a catheter placed in the arterial blood vessel and an associated system consisting of tubing, pressure transducer and an electronic processor. Included in the processor are filters, amplifiers and readout devices. An accurate measurement of the pressure waveform requires an understanding of the individual pressure monitoring system and the potential errors that the system is prone. Certain generic errors which can be made when using pressure monitoring systems include zeroing, calibration and dynamic response. In invasive, or intra-arterial BP (IABP), the measurement of MAP (mean arterial pressure) may be more reliable than systolic or diastolic pressures, as MAP is less prone to error caused by waveform damping, artifacts (Fig. 7.3).

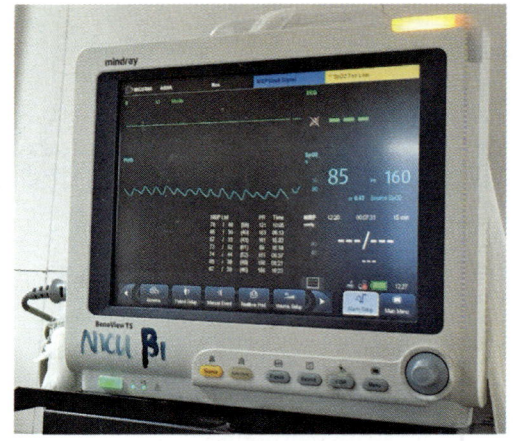

Fig. 7.3: A multipara monitor with facility of displaying HR, saturation BP (invasive and noninvasive)

For invasive measurement accuracy
1. Zero, all electronic systems internally.
2. Maintain transducers at the level of the atria.
3. Calibrate transducer daily with mercury.
4. The connecting tube between the arterial catheter and the pressure transducer affects dynamic response. This tubing should be of the shortest possible length.
5. Tubing should have low compliance.
6. No more than one leur-lock connection should be used between the pressure transducer and the arterial catheter.
7. No air bubbles should be present in the system for they can result in over damping of the signal, leading to inaccurate measurement.
8. A long and compliant tubing can result in under damping of the signal and ringing of the waveform leading to overshooting of SBP and undershooting of DBP.

All connections should be tightly secured and leur-locked. Invasive blood pressure gives an accurate measurement but its disadvantages are that it is very costly and is associated with all the complications of an invasive line (Table 7.3).

TABLE 7.3: Common brands of blood pressure monitors available in market

S.No.	Make	Dealers	Options*	Unit cost (₹)
1.	Kenz BPM OS-22 CAS NIBP monitor	Global Med System	1,2	1,50,000
2.	PM-600 I	Medical Systems	1,2	60,000
	PM 600 II		1,2,5	1,20,000
	PM 600 III		2,5	50,000
3.	*In vivo* Omega 1440	Omega Med Pvt. Ltd	1,2	56,000
	1445		1,2,3,5,6	3,00,000
4.	Monitor APM 1000	Global Med System Model	1,2	1,50,000
	740		1,2	1,10,000

(Contd...)

Blood Pressure Monitors

TABLE 7.3: Common brands of blood pressure monitors available in market (Contd...)

S.No.	Make	Dealers	Options*	Unit cost (₹)
5.	Pressmate BP 8800	Hospimedica Int	1,2	1,05,000
6.	Ivy 403 Ivy 405	Rustagi Surgicals	4,3,6,8 1,2,3,4,5,6,8	3,50,000 4,50,000
7.	Cardiocap II	AVL Biomedicals	1,2,3	1,70,000
8.	Horizon 1100	Omron Med Pvt Ltd	1,2,3,4,5,6	8,00,000
9.	Pacetech Vitalmax 800 Vitalmax 4000B Vitalmax 4100G-2(P)	Medex Agencies	 1,2,5 1,2,3,5,6,7,8 1,2,3,4,5,6,7,8	 1,80,000 4,00,000 5,00,000
10.	Criticare 503, 507	Criticare India Ltd.	1,2,5	3,00,000
11.	Orbit-n	Larsen and Toubro Ltd.	1,2	1,00,000
12.	Nihon Koden	Nihon Koden	1,2,3,4 5,6,8	5,00,000
13.	Gabriel	Vishal Surgical	1,2,3,5,6,7,8	5,50,000
14.	Genera 200	Intermedical Inc	1,2	80,000
15.	Dianamap pro100	Datex Ohmeda	1,2	1,00,000
16.	EMCO	EMCO Meditek	1,2	70,000
17.	SureSigns VM 6 (NIBP)	Philips	1,2,3,4,5,6 8 is optional	200,000
18.	Mindray VS-800 BP monitor	Rohanika/Mindray	1,2,5,8	100,000
19.	Schiller BP-200 monitor	Schiller India/Germany	1,2,5,8	300,000
20.	Minitorr Plus NIBP monitor	DOT Med/Smith Medical	1,2,5,8	100,000

Options:
1. NIBP 2. HR 3. ECG 4. IBP
5. SaO2 6. RR 7. EtCO$_2$ 8. Temp

Frequently Asked Questions (FAQs)

Q. 1. What minimal displays should be there on an automatic noninvasive monitor?

Display should include systolic, diastolic, mean blood pressure and heart rate values. Audible and visual alarms should be inbuilt. There should be a provision to record at regular intervals.

Q. 2. What are the indications for recording invasive blood pressure?

1. In very small or unstable infants, particularly those with severe hypotension.
2. During major procedures that could cause or exacerbate intravascular instability.
3. To monitor infants on aggressive ventilator support.

Q. 3. What are the additional advantages of recording invasive blood pressure?

Invasive blood pressure provides a continuous documentation of mean arterial blood pressure (which approximates nearly to the gestational age in weeks). The pulse pressure of >20 mm of Hg (difference of systolic and diastolic) provides a guide that ductus arteriosus may be open in a preterm baby. In addition, it provides an easy site for taking blood samples for investigations.

Q. 4. When should one resort to invasive blood pressure monitoring?

Once a unit is providing reasonable level II care and venturing for intensive care (including ventilation). Remember that one has to use continuous infusion of heparinized saline through the line with the help of infusion pumps.

	Normal BP readings	
Weight	Systolic	Diastolic
1000 gm	36–59	17–38
2000 gm	42–65	21–41
3000 gm	50–72	27–46

(From Avery 1994; pp. 1405)

Q. 5. What are approximate normal blood pressure readings in neonates?

Normal mean BP

Lower limit of mean BP = gestational age of baby in weeks

Upper limit of mean BP = gestational age of baby in weeks + 20

Q. 6. What is the standard protocol necessary for oscillometric blood pressure measurement in term newborns?

For routine care of term newborns, blood pressure measurements with the oscillometric technique may be made without the need of a special position or sleep state, provided that the measurements are made with an appropriate sized cuff in the absence of struggling, crying and movement of the newborn.

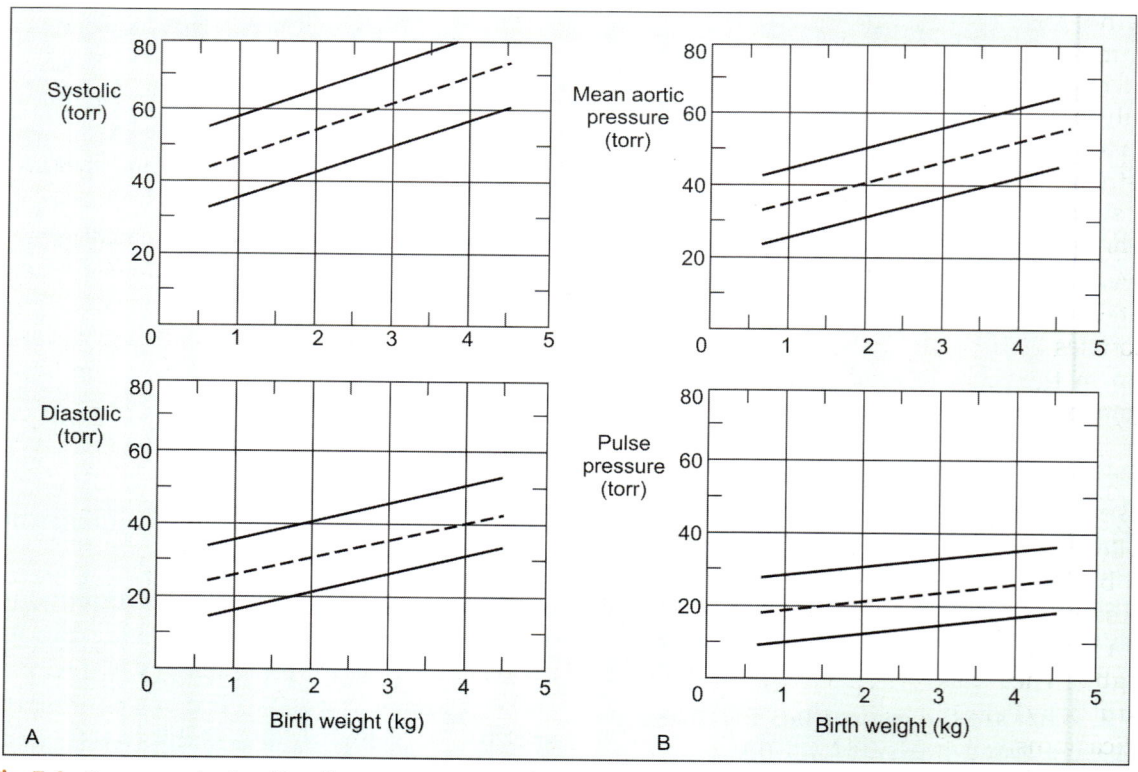

Fig. 7.4: Pressures obtained by direct measurement through umbilical artery catheter in healthy newborn infants during first 12 hours of life. Broken lines represent linear regressions; solid lines represent 95% confidence limits. (A) Systolic pressure (top) and diastolic pressure (bottom). (B) Mean aortic pressure (top) and pulse pressure (systolic-diastolic pressure amplitude) (bottom)

Chapter 8

Transcutaneous Bilirubinometer

Neonatal jaundice occurs in nearly 70% of term and 80% of preterm babies. Management of jaundiced neonates often requires estimation of total serum bilirubin (TSB). Lab estimation of TSB is commonly based on van den Bergh reaction. There is marked interlaboratory variability. Micro-methods for bilirubin assay, based on twin-beam spectrophotometry, are more accurate and require lesser quantity of blood. However, both the methods require drawing of blood causing pain and trauma to the neonate and distress for parents. These problems have led to search for non-invasive and reliable technique for estimation of TSB.

A large number of studies have demonstrated the prediction of serum bilirubin in neonates by measuring the yellowness of the skin in the jaundiced neonate using transcutaneous bilirubinometers.

Principle

High correlation between cutaneous bilirubin (yellowish discoloration of skin) and TSB form the basis of transcutaneous bilirubinometry. These bilimeters are based upon reflectance data analyzed on dual or multiple wavelengths. They measure spectral reflectance of bilirubin by determining the difference between optical densities of light. These optical signals are converted to electrical signal by a photocell, which are later analyzed by a microprocessor to generate a transcutaneous bilirubin (TcB) value in mg/dl or μmol/L.

The major skin components which impart the spectral reflectance in neonate are:
 i. Melanin
 ii. Dermal maturity
 iii. Hemoglobin
 iv. Bilirubin

The available meters can be divided into two categories
 i. Multi-wavelength spectral reflectance meters (BiliChek)
 ii. Dual wavelengths (450 nm, 550 nm) spectral reflectance meters (JM-103)

JM-103 (Minolta Airshields) measures the spectral reflectance of bilirubin by determining the difference between optical densities for light in the blue (450 nm) and green (550 nm) wavelengths—a dual optical path system. The measurement of bilirubin accumulated primarily in the deeper subcutaneous tissue should decrease the influence of other pigments in the skin, such as melanin and hemoglobin.

The BiliChek system (SpectR$_X$, Norcross, GA) performs a spectral analysis at more than 100 different wavelengths. By subtracting the spectral contribution of the known components, the bilirubin absorbance is quantified.

How do these meters report the results?

The earlier transcutaneous bilirubinometers report the results in form of transcutaneous bilirubin index (TcBI). The TcBI can be converted to bilirubin values in mg/dl or µmol/L by using different multiplication factors for different populations. However, both the currently available bilimeters display the results in clinically appropriate units: mg/dl or µmol/L.

Basic Operating Procedure

While each transcutaneous bilirubinometer has a different detailed operating procedure, the basic principle remains the same. The optic head of the meter is placed against the forehead or sternum of the neonate and gently pressed. For correct measurement, the optic head should make full contact with the skin and there should be no gaps between the optic head and the skin. This can be achieved by gentle pressure. While JM-103 give TcB value based on single measurement or average of three repeated measurements, BiliChek require five replicate measurements at one site and give the TcB value based on average of the five replicate measures (Fig. 8.1).

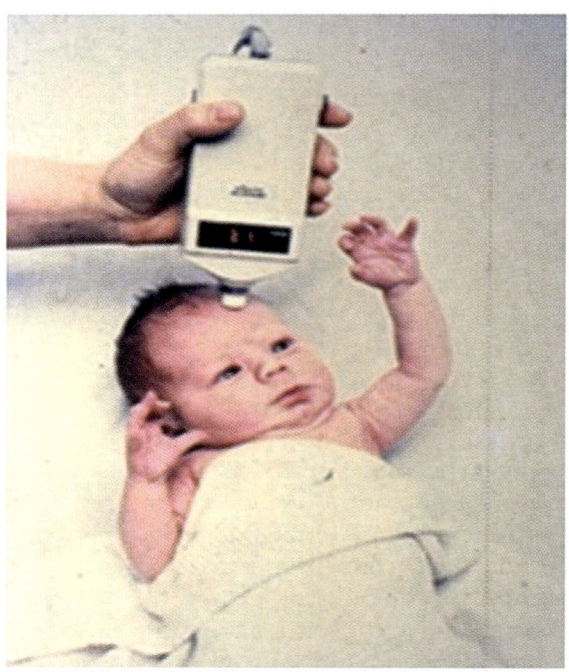

Fig. 8.1: Transcutaneous bilimeter with direct display of bilirubin in mg/dl

Site of Measurement

The recommended sites for TcB assessment by any bilimeter are the forehead and the upper end of sternum. Hyperemia at the test site may affect the results. Measurements against bruises, birthmarks and subcutaneous hematoma should be avoided.

Clinical Utility of Transcutaneous Bilirubinometers

Studies have demonstrated good correlation between transcutaneous bilirubinometer results and measured TSB. Most studies have found fair to excellent correlation and agreement between the two. As per American Academy of Pediatrician 2004 guidelines on management of hyperbilirubinemia in term and near-term neonates, transcutaneous bilirubinometry can be used as a surrogate of STB for screening of hyperbilirubinemia. However, bilirubin levels need to be confirmed with measured TSB before starting any intervention for hyperbilirubinemia (Table 8.1).

TABLE 8.1: Common transcutaneous bilirubinometer available in the market

S.No.	Make	Dealer's name	Unit cost (₹)
1.	JM-103 Jaundice meter Konica/Minolta	Hillrom-Airsheild Draeger Medical	1,00,000
2.	BiliChek	Rustagi Surgicals Phoenix Medical	1,75,000
3.	Bilitest	Lacteromedik	1,20,000
4.	KJ-8000, China	Medexcel, New Delhi	80,000
5.	Bili-2000 jaundice detector	Kruise Pathline Private Limited	1,00,000

Frequently Asked Questions (FAQs)

Q. 1. Why is there a need for transcutaneous bilirubinometry?

Neonatal jaundice is a very common condition. Proper assessment and appropriate management of hyperbilirubinemia is very important to prevent kernicterus, a rare but devastating complication of exaggerated hyperbilirubinemia. TSB measurement in clinical laboratory is an objective method, but it is expensive, invasive and there is significant inter-laboratory and intralaboratory variability. The necessary blood sampling is painful and associated with the possibility of infection. Visual assessment of bilirubin is observer-dependent and therefore less reliable. Transcutaneous measurement of bilirubin concentration is considered to be ideal for neonatal jaundice screening. Transcutaneous readings are immediate, and they can indicate the need for TSB testing. Routine use of transcutaneous bilirubinometry compared to visual assessment of bilirubin is associated with reduction need for blood sampling.

Q. 2. What is the principle of transcutaneous bilirubinometry?

High degree of correlation between cutaneous bilirubin and TSB is the basis of transcutaneous bilirubinometry. Simply stated, transcutaneous bilirubinometers measure yellowness of the skin by analysis of spectrum of light (including dual or multiple wavelengths) reflected by the baby's skin.

Q. 3. How newer transcutaneous bilimeters are different from older one?

Newer transcutaneous bilirubin measuring devices (BiliChek and JM-103) are compact and lighter weight. Performance of newer devices is much better compared to previous ones. These devices estimate serum bilirubin in clinically useful units like µmol/L or mg/dl.

Q. 4. What are the differences between BiliChek and JM-103?

The Konica Minolta/Air-Shields® **JM-103 Jaundice** Meter® from Draeger Medical is an accurate, instantaneous, non-invasive and a dual optical path system device that provides an estimate of serum bilirubin levels reported in µmol/L or mg/dl. To use the unit, simply place the tip of the device against the forehead or sternum of the infant and press gently. The instrument's xenon lamp will flash once and the measured value will be displayed on the screen. No additional data analysis is required by the user and the unit may be used with all skin colors. The JM-103 Jaundice Meter® requires no disposable parts.

The *BiliChek* System® (SpectR$_X$, Norcross, GA) is a multi-wavelength reflectance device which performs a spectral analysis at more than 100 different wavelengths. By subtracting the spectral contribution of the known components, the bilirubin absorbance is quantified and an estimate of serum bilirubin level in µmol/L or mg/dl is performed. Each time before use device need to be calibrated with a disposable BiliCal® individual calibration tip. Though BiliCal® eliminates the possibility of cross-infection by providing a new, clean surface for every patient, it increases ongoing cost of the instrument. To use the unit, simply place the tip of the device against the forehead or sternum of the infant and press gently. Once the instrument's xenon lamp flashes green, press the trigger to take the reading. Five repeat measures/readings need to be done. Device calculates the average of the five measured values and display on the screen.

Q. 5. What are the limitations of transcutaneous bilirubinometry?

Transcutaneous bilirubinometry is definitely more accurate compared to visual assessment of bilirubin and being a non-invasive method it is very useful device for screening of hyperbilirubinemia. However, at higher bilirubin ranges (>15 mg/dl) it significantly underestimate the TSB and thus have limited value. Secondly, for up to 5% of neonates values of transcutaneous bilirubinometry may be grossly abnormal. Clinical assessment of bilirubin along with transcutaneous bilirubinometry to decide the need for blood sampling to confirm TSB may safeguard against these errors.

Q. 6. What are the commonly used sites for estimation by these meters?

Forehead and the upper end of sternum are the commonly used sites. Hyperemia at the test site may affect results. Measurement against bruises, birthmark and subcutaneous hematoma should be avoided.

Q. 7. Does it matter who does the measurement?

If the instrument is rightly used, it does not matter who does the measurement. This alleviates the need of pediatrician/skilled health care provider for screening the babies for hyperbilirubinemia. However, whenever possible clinical assessment should also be made to safeguard against infrequent but grossly erroneous readings of these devices.

Q. 8. Do these instruments have any ongoing costs?

For BiliChek®, the disposable calibration tip (BiliCal) needs to be replaced for every new measurement, which adds to the cost of operation. There is no ongoing cost for JM-103.

Q. 9. How sensitive and specific are the bilirubin estimations by transcutaneous bilirubinometer?

Transcutaneous bilirubin (TcB) measured by multi-wavelength spectrophotometry has positive and a linear correlation with STB, and serves as a useful tool to quantify jaundice. Devices such as BiliCheck® claim to correct for the confounding effect of skin pigmentation and are reported to have closer agreement with STB (Pediatrics 2004; 114; 130–153).

Minolta JM-103 TcB assessment also demonstrates significant accuracy compared to TSB and can be used as a screening test to identify the need for blood sampling for TSB levels in term and near-term newborn infants (Pediatrics 2004;113:1628–1635). Both the devices have demonstrated significant accuracy in preterm neonates as well.

Q. 10. What is the clinical utility of transcutaneous bilirubinometer? Are these useful as screening tools?

Presently available instruments are not sufficiently sensitive and specific to replace TSB measurement. Though these devices are reasonable accurate at lower TSB levels, at higher TSB levels the difference between TcB estimate and TSB appears to be too large to be acceptable. However, transcutaneous bilirubinometers are useful screening tools for deciding the need for blood sampling to assay TSB.

Chapter 9

Thermometers

Newborn infants grow better if their core body temperature stays in normal range (36.5 to 37.5°C). A rise or drop in body temperature increases the metabolism resulting in both calorie and oxygen consumption.

Temperature measurement is a very important part of monitoring of a neonate. A variety of devices are available for measuring the temperature of a newborn and these use different principles to arrive at the temperature value.

Temperature monitoring can be intermittent or continuous, and based on this, the instruments used vary.

INTERMITTENT TEMPERATURE MONITORING

A. Instruments Available

1. Mercury in Glass Thermometer

This is the standard thermometer used and is readily available. The instrument consists of a glass reservoir with mercury connected to a glass stem with temperature markings. Upon contact with the body, the mercury in the bulb expands into the glass stem and the temperature can be read off. Also known as clinical thermometer, these instruments have a kink in the glass stem at the junction of the mercury reservoir and so that the mercury does not fall back into the reservoir once the thermometer is removed from the body. After reading, the instrument is shaken to enable the expanded mercury to go back into the reservoir. Rectal temperature in neonates is no longer measured routinely because there is a serious risk of rectal perforation associated with the procedure. Axillary temperature is measured and correlates fairly well with the core temperature.

For documenting hypothermia, a low-reading thermometer is needed to measure the subnormal temperatures. For measuring rectal temperature, a rectal mercury thermometer is used. Ordinary clinical thermometers are not be used for rectal measurements because of the risk of perforation. Rectal thermometer has a short and thick bulb.

Due to risk of environmental mercury toxicity, these thermometers are being phased out.

2. Digital Electronic Thermometers

Here the sensor is either a thermocouple or a thermistor probe and when placed in contact with the body, the temperature sensed by the probe is electronically processed and the reading displayed digitally.

3. Infrared Thermometry

Also known as non-contact thermometry or tympanic thermometry; an infrared sensor placed in the external ear canal detects infrared radiation from the tympanic membrane and displays the temperature based on this. This reading is more close to the core temperature because the tympanic membrane shares its blood supply with the hypothalamus. This method takes very little time to record—just 1 to 2 seconds. This technique may not be feasible in preterm infants.

B. Method of Recording Temperature

1. Mercury Thermometers

a. Shake the thermometer vigorously to get the mercury down
b. Ensure that the axilla is dry
c. Place the bulb of the thermometer at the apex of the axilla and the stem parallel to the lateral chest wall
d. Hold the infant's arm firmly against the chest wall to keep the thermometer in place
e. Hold in place for at least three minutes and then read off the temperature.

2. Digital Thermometers

a. Follow the same method as for mercury thermometer
b. Hold the instrument in place till a beep is heard, indicating that the measurement process is over
c. Read the temperature off the panel.

3. Tympanic Thermometers

a. Apply disposable cover to the sensor head
b. Ensure that measurement is made in the ear canal that was uncovered/unobstructed for at least 5 minutes before the reading (otherwise a falsely high reading may be obtained)
c. Gently insert the probe in the ear canal; *do not force the probe*
d. Press the measurement button once the probe is in place and steady
e. Remove the probe from the ear canal and read the temperature
f. Discard disposable cover.

C. Advantages and Disadvantages of Each Type of Instrument

1. Mercury Thermometer

a. *Advantages*
 i. Easily available and cheap
 ii. Readings fairly accurately approximate core temperature

b. *Disadvantages*
 i. Prone to breakage
 ii. If the markings on the instrument are not proper, reading off the temperature may become difficult
 iii. Thermometer has to be held in place for the entire duration of measurement, which is a long duration of at least three minutes
 iv. The accuracy of commercially available mercury thermometers has to be cross-checked periodically using either water baths whose temperature is measured with a standardized and accurate thermometer or with heat blocks of known temperature.

2. Digital Thermometer (Fig. 9.1)

a. *Advantages*
 i. Shorter duration of measurement
 ii. Gives reading to the nearest 0.1°C

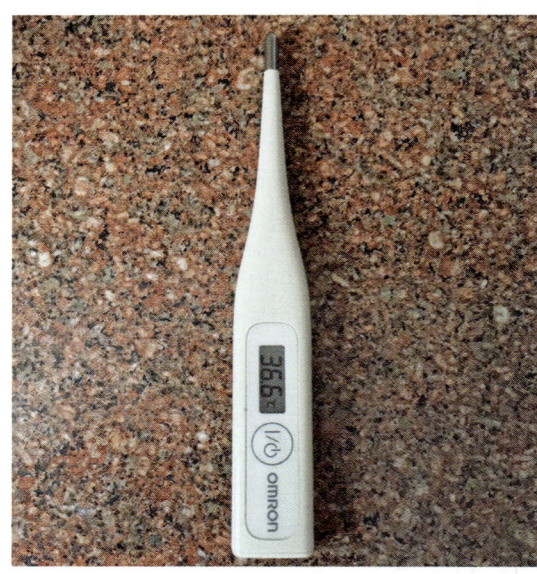

Fig. 9.1: Digital thermometer

iii. Gives audible signal at the end of measurement
b. *Disadvantages*: Expensive.

3. Tympanic Thermometer

a. *Advantages*
   i. Non-contact method and use of disposable sensor covers prevents transmission of infections
   ii. Short time (1 to 2 sec) taken for temperature determination
   iii. Measurement approximates core temperature closely.

b. *Disadvantages*
   i. Very costly equipment
   ii. Sensor size precludes its use in very small infants
   iii. Cannot be used in infants with middle ear disease or those with malformations of the external ear canal
   iv. Forcing the probe into the ear can cause damage to the tympanic membrane.

Continuous Temperature Monitoring

A. Need for Continuous Temperature Monitoring

Continuous monitoring of temperature is needed in the following situations:
   i. A very sick neonate where the trend of temperature is to be studied
   ii. A neonate with hypothermia, when being re-warmed
   iii. Continuous temperature is measured to provide feedback to servo-controlled heating devices providing a thermo-neutral environment to the infant.

B. Probes Available

1. ***Thermistor probes***: Thermistor probes have a resistive element that is temperature sensitive. The resistance decreases in proportion to the temperature increase. This change in resistance proportionately changes the current flowing through the probe and the current flow is sensed electronically and displayed as temperature.

2. ***Thermocouple probes***: This probe looks like a small bead and has two different metals that generate a very small voltage that is proportionate to the temperature sensed. The voltage generated by the probe is sensed electronically and converted to temperature reading.

C. Method of Applying Probe

1. Ensure that the probe and the skin surface where the probe is to be applied are clean.
2. The probe is to be applied on the skin over the upper part of the anterior abdominal wall in infants who are supine and on the skin of the flank of infants who are being nursed prone.
3. Fix the probe with micro-pore or other similar material.
4. Cover the probe with a reflective pad to reflect away additional heat from other devices like phototherapy units.
5. Check periodically to ensure that the probe is properly positioned and in contact with the skin of the infant.

D. Precautions to be Taken during Continuous Temperature Monitoring

1. Do not apply probe to breached or bruised skin surface.
2. Do not apply probes over dressings.
3. Ensure that the probe is not in contact with the bed or the surface on which the infant is being nursed. The probe should not be sandwiched between the mattress and the skin, an artificially high skin temperature will be recorded and this will result in an inadvertent decrease in heater power output.
4. Ensure that the correct surface of the probe is in contact with the infant's skin.
5. Change the site of the probe frequently.

TABLE 9.1: Common thermometers available in Indian market

S.No.	Make	Principals	Dealers	Unit cost (₹)
1.	Mercury	Hicks	Local	50/-
2.	Electronic* digital	Becton-Dickinson	B and D	350/-
3.	Omron digital	Omron	A and R	350/-
4.	Infrared tympanic	B Braun	B Braun	1750/-
5.	Infrared skin	Chicco	—	3500/-
6.	Electronic probe	Fisher–Paykel Alaris Medical	Fisher–Paykel Medex	5,000/-

*Available in open market—China, Taiwan and Korean make.

6. Cross-check the temperature indicated by the probe with manual measurement by another thermometer at least two hourly; especially when the continuous temperature monitoring is being done for servo-controlled devices.

E. Warning!

Though the thermistor and thermocouple probes may look similar, they are not interchangeable and they work using different principles. Therefore, make sure you are using the correct probe as specified by your machine manufacturer.

Frequently Asked Questions (FAQs)

Q. 1. What special precautions should be observed for using a thermometer?

1. Each infant should have a separate thermometer.
2. Disinfect thermometer with alcohol after use, keep thermometer dry and not in the disinfectant solution.
3. Do not use clinical glass thermometer for rectal temperature recording. Rectal thermometer has short and thick bulb.

Q. 2. Which thermometer is preferable, digital or mercury?

Digital thermometers are now being used instead of the traditional mercury-in-glass thermometer. This thermometer has a thermistor at the tip that measures the peak temperature reached in the surrounding tissue and converts it to a digital display. Most published literature that supports the use of this type of thermometer is beyond the neonatal period. There are only a few studies where neonates, especially in preterm infants, were the selected patient group. The accuracy and precision of the instrument is approximately ± 0.2°C.

(India will have to phase out mercury by 2020 as the country has signed a global treaty—Minamata Convention—which makes it mandatory for the signatories to ban the use of the deadly nerve toxin in a phased manner)

Q. 3. Enumerate characteristics of a good neonatal temperature monitoring device.

1. The instrument should have a resolution of 0.10°C.
2. It should be capable of displaying the temperature in both centigrade as well as fahrenheit.
3. The instrument should be accurate and there should be a provision to standardize it.
4. Measurement of temperature should be easy and fast.
5. The instrument should be easy to clean and sturdy as well as reasonably priced.

Q. 4. What are newer temperature measuring devices?

There have been a number of innovations keeping in mind the qualities desired in a good thermometer and a number of new instruments are available in the foreign market.

Some of them are available in the Indian market too.

It should be stressed that all the devices described below are not routinely recommended for temperature monitoring in neonates. Some of these devices have not been studied in neonates and some have been found to be inaccurate. Till there is scientific evidence regarding the accuracy of these devices for temperature measurement in neonates, these are not recommended for use in this group of patients.

1. *LCD skin thermometer*: This device is applied to the skin, usually on the forehead and the temperature is displayed digitally by the liquid crystal diode in the device. Though easy to use, it is not very accurate for use in newborns.

2. *Pacifier thermometer*: This device has a thermometer incorporated into a pacifier. After turning on the device, the pacifier is placed in the infant's mouth when the device gives a beeping sound to indicate that it is ready to measure. It measures temperature with a resolution of 0.2°F and has a battery that lasts for 18 months. Apart from the obvious difficulty in cleaning the device for multiple patient use, this device has not been tested on infants.

3. *Infrared sensing skin thermometers*: Using a principle similar to the tympanic thermometer, this device senses the radiant energy from the skin and reads it off as temperature. Needs validation before being recommended for use in neonates.

Chapter 10

Weighing Scales

Weight record is essential to monitor the adequacy of nutrition as well as fluid balance. Accurate weighing scale is a fundamental need for all special care neonatal units and delivery rooms. Recording weight at birth and daily is essential for the management of very low birth weight (VLBW) infants. Weight of birth is the singlemost useful predictor of neonatal morbidity and mortality. Birth weight helps in identifying the level of care required for the infant and classification into weight-for-date categories. Infants below 2000 gm have special needs and need nursery care. Small-for-dates and large-for-dates infants also need special newborn care. Hence, a weighing scale for measuring the weight at birth is essential for all facilities where deliveries take place and where neonates are looked after.

Indications
- All infants at birth.
- All LBW infants at 2 weeks (to check regaining of the birth weight), 4 weeks (to ascertain a weight gain of 80–100 gm/kg per week) and then every month.
- Sick newborn once or twice a day.
- VLBW (<1500 gm) infants once or twice daily to monitor fluid therapy.
- Measuring urine output by pre-weighed napkin.

Sick and VLBW infants need daily weighing to decide fluid requirements, drug dosages and weight gain patterns. Accurate daily weighing would be helpful in avoiding complications due to under or over hydration. Excessive weight gain would raise suspicion of fluid overload or of congestive cardiac failure/acute renal failure. Sudden weight loss in an infant who had been gaining weight satisfactorily suggests the possibility of dehydration. Adequate daily weight gain in a newborn is a sensitive index of its well-being. Term infants lose about 10% of birth weight and regain birth weight at 7 to 10 days of age while preterm infants lose weight during the first 3–4 days of life and can lose up to 10–15% of the birth weight and regain birth weight usually by 14–21 days of age. After the initial weight loss, infants start gaining weight at a rate of 1–1.5% of birth weight per day. Scales with an accuracy of ±5 gm are essential in the weight of monitoring of VLBW infants. Newborn units that manage infants under 1000 gm in weight need weighing scale with accuracy of 1 gm. Excessive weight loss, delay in regaining birth weight or slow weight gain suggest that either the infant is not being fed adequately or the newborn is unwell and needs attention.

A weighing scale can be employed to measure the urine output of the infants. Pre-weighed nappies should be used for nursing infants. Weighing the nappies post-voiding would be helpful in assessing the urine output of sick infants. Weighing a infant pre- and post-feed is helpful in assessing adequacy of feeding in breastfed newborns.

Desirable Specifications

- Table top, light and portable
- Built in rechargeable battery
- Hygienic, easy to clean infant tray
- Acrylic (nonmetallic) infant tray
- Reproducible weights
- Resolution of ± 5 gm (optional ± 1 gm)
- Freeze reading display
- Zero weight adjustment facility
- Quick and clear digital read outs
- Measurement does not change with position of infant on the pan.

Types of Weighing Scales

1. Spring balance
2. Electronic weighing scales (preferable)

In all weighing scales there is a weight sensitive device attached beneath the infant pan, a spring for a conventional spring balance weighing scale and an electronic sensor for an electronic weighing scale. For most scales the infant should be placed in the center of the pan for an accurate reading and the reading may vary with the position of the infant on the pan. Newer electronic scales have tried to overcome this disadvantage of position variation by using a special type of electronic weight measuring device (called load cell). In weighing scales with this facility the infant may be placed anywhere on the infant pan and the reading will always be the same.

Electronic weighing scales are preferable to spring balance scales because of greater: Reproducibility, reliability, resolution. They also provide quick digital readout (Fig. 10.1 and Table 10.1).

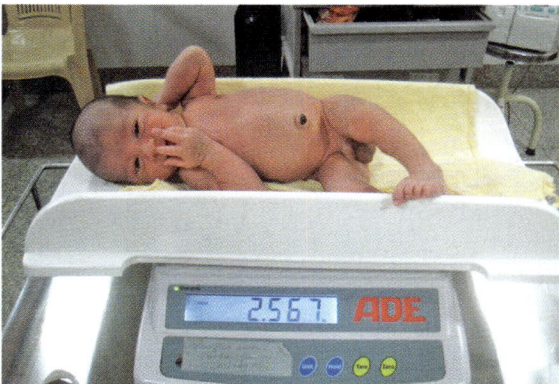

Fig. 10.1: Electronic weighing machine with sensitivity 1 gm

TABLE 10.1: Weighing scales available in Indian market

S.No.	Model, capacity, resolution	Prinicipals	Dealers	Unit cost (₹)
1.	Model 1, 0–20 kg, ±5–10 gm Model II, online, ±1 gm	Phoenix, Zeal	Phoenix	12,500/- 14,000/-
2.	NICU model MI 20, 0–20 kg, ±5 g MI 5, 0–5 kg, ± 1 gm	Zeal, Meditrin	Zeal, Meditrin	14,000/- 14,000/-
3.	Scale for weighing urine MU 500, 0–500 g, ±1 gm	Meditrin	Meditrin	8,500/-
4.	Seca 727/728, 0–20 kg, ±2 gm Seca 736, 0–15 kg, ±5 gm Seca 734, 0–20 kg, ±10 gm	Seca	Scorpia India Ltd.	30,000/- 26,000/- 12,000/-
5.	Manual 10, 0–20 kg, ±10 gm Auto 5, 0–10 kg, ±5 gm	Lectromedik Shreeyash Electo Medicals	Lectromedik Shreeyash Electo Medicals	13,000/- 15,000/-
6.	Warm weigh, 0–10 kg, ±1 gm	Air-Shield-Hill-Rom Healthcare	Wipro GE	60,000/-
7.	ADE, 0–10 kg, ±1 gm	Rustagi Surgicals	ADE Germany	80,000/-
8.	Venus digital electronic scale EBS-9090	Ace incorporation		
9.	CTL 208, CTL 3000	Cardiocare	Citizen	

*Avery India Ltd., Osaw Industrial Pvt. Ltd. also manufacture good weighing scales.

Procedure

1. Put the weighing scale on a flat, stable surface.
2. Record weight prior to feeding.
3. Detach as many tubes/equipment as possible. Keep the naked infant on the towel and record the weight (subtract the weight of the towel if the scale has no facility to zero).
4. Keep infant in middle of scale pan; hold the remaining tubes and lines in hand.
5. Use separate sterile towel for each infant.
6. If using pre-weighed splint, reduce the weight from infant's weight.
7. For quality assurance check accuracy of weighing scale with standard known weights every week.

Operating Instructions

1. The weighing pan should be cleaned before weighing each infant.
2. Connect to the mains and switch on the machine.
3. The digital display will show some figure.
4. Place a sterile towel or paper on the pan to reduce the chances of hypothermia and cross infection.
5. Adjust the digital display to zero by manually adjusting the knob. Some weighing scales have automatic zero facility.
6. Place the infant on the towel/ paper, in the middle of the pan.
7. Note the reading on the digital display. Freeze reading facility will continue to show the reading even after the infant is removed from the scales.
8. The machine should be switched off after use.
9. Do not press the weighing pan with your hand. It could damage the load cell system in the weighing machine.

Frequently Asked Questions (FAQs)

Q. 1. How often should a weighing scale be calibrated?
A weighing scale should be calibrated at least once a week. Calibration can be checked in the unit against known standard weights, e.g. 1/2/3 kg.

Q. 2. What is an online weight measurement system?
An online weight measuring system is the latest modification in weighing machines. It consists of two components, a weighing plate connected with a cable to the display unit. The plate could be placed under the infant while nursed in the open care warmer or incubator. It is also X-ray cassette compatible. The display unit is a separate module and can be attached at a convenient location for better visibility. This is particularly useful for weighing sick infants with feeding tubes, electrodes, IV lines and endotracheal tubes who are connected to the ventilator. Since the infant does not have to be moved for taking the weight, the disturbance to the infant is minimal and the convenience to the nursing staff increases.

Q. 3. What weight range should be kept in mind when selecting a weighing scale?
It depends on the unit. If the weighing scale is to be used for the newborn unit exclusively, machines with a range of up to 5–7 kg with a resolution of ± 1 gm would be ideal. This system could also be used to assess the urine output by using pre-weighed nappies. However, if the system is to be used in the neonatal unit as well as in the neonatal follow-up clinic, machines with a range of up to 10–20 kg with a resolution of ±5–10 gm can be selected.

Q. 4. How should the infant pan be cleaned before use?
It is very important to clean the infant pan before and after weighing each infant. A single weighing scale in the unit could be a source

of infection. Commonly available disinfectants like savlon, cidex may be used to clean the pan. Spirit/alcohol should be avoided as it can damage the pan material or LED display. If the infant pan is detachable major stains like blood and stools can be cleaned with a detergent and water. Further a sterile towel/paper can be placed on the pan before weighing the infant which should then be changed before weighing each infant.

Q. 5. Where a weighing scale should be kept in the unit?

The weighing scale could be placed on any mobile trolley and should have an inbuilt rechargeable battery. The machine is often required to weigh VLBW/sick infants on the ventilator, in incubators or open care warmers. A mobile weighing machine which could be wheeled in near to the newborns bed would be convenient in weighing such sick infants.

Chapter 11

Transcutaneous Blood Gas Monitors

Transcutaneous measurement of PO_2 ($TcPO_2$) and pCO_2 ($TcPCO_2$) provides instant and continuous information on the body's ability to deliver oxygen to the tissues and remove carbon dioxide via the cardio-pulmonary system. The electrodes measure the gas tension of the underlying tissues and not the arterial gas tension. When hemodynamic conditions are stable, transcutaneous measurements correlate well with arterial values but this does not necessarily mean that they are identical. As against the one point information provided by blood gas estimation, transcutaneous monitoring gives a real time monitoring.

Transcutaneous gas monitors were used widely in NICUs in 1970s and early 1980s. With the easy availability of compact and handy pulse oximeters in the 1980s, these have almost completely replaced $TcPO_2$ monitors in most NICUs. However, $TcPO_2$ monitors have also undergone improvement in design and reduction in size over the recent years and still offer some benefits for babies who are extremely low birth weight (<28 wk, <1 kg) during initial days of life. Currently both $TcPO_2$ and $TcPCO_2$ monitors are available as one combined solid state electrode.

ADVANTAGES OVER PULSE OXIMETER

a. The driving force for cellular oxygen uptake is PO_2 in the tissues and $TcPO_2$ directly measures it, whereas pulse oximeter gives information only percentage of oxygen saturation of hemoglobin. By using oxygen dissociation curve, PO_2 can be derived from oxygen saturation. But this relationship is quite variable in a neonate and is influenced by factors such as pH, pCO_2, temperature, 2, 3-DPG and HbF.

b. Pulse oximeter gives no information on pCO_2 status which is a measure of the adequacy of ventilation and also influences the cerebral and peripheral circulation.

c. Pulse oximeter cannot detect hyperoxia reliably, whereas $TcPO_2$ monitor gives the actual PO_2 values.

PRACTICAL WORKING

The $TcPO_2$ monitor consists of an electrode which is applied to the skin and heated electronically to 43° to 43.5°C. This electrode is plugged into an amplification unit which displays digital read out of oxygen tension ($TcPO_2$). The $TcPO_2$ electrode is about 1.5 cm in diameter and consists of a ring-shaped silver anode, a central gold or platinum cathode, heating element and thermistor, all encased in plastic. The working surface of the electrode is covered with a polypropylene or teflon membrane which is separated from the electrode by electrolyte solution. Most electrodes have either a pre-packed snap or screw on type membrane assembly with retainer rings. The electrode is calibrated against a null solution or by electronic zeroing (low) and in room air or gas of known oxygen concentration (high).

Site of Application
The electrode should be applied to an area with homogenous capillary bed without large veins, skin defects, hair or bone. The anterior abdominal wall offers a convenient site.

Principle
The heating of the electrode increases capillary blood flow and the partial pressure of oxygen and carbon dioxide and makes the skin more permeable to gas diffusion. The oxygen reaching the skin surface diffuses through the membrane of the oxygen sensor into the electrolyte solution. On application of an appropriate polarization voltage, the oxygen is reduced at the cathode and current flows between cathode and anode in direct proportion to the PO_2.

Correlation of $TcPO_2$ and PaO_2
Since metabolism in tissue consumes oxygen and produces carbon dioxide, transcutaneous values differ from arterial values. Typically PaO_2 is a little lower than in the arterial blood and $PaCO_2$ is a little higher when measured transcutaneously.

$TcpO_2$ provides information on oxygen supply to the tissues and is dependent on the oxygen uptake of the respiratory system, the oxygen transporting capacity of blood and condition of the circulatory system. If a baby is hemodynamically stable, $TcPO_2$ correlates well with PaO_2. Under such circumstances, transcutaneous monitoring continuously provides information on the baby's respiratory status. If changes take place in the blood's capacity to transport or release oxygen, $TcPO_2$ will reflect the influence of this event on the oxygen supply to the tissues. If respiration is stable but baby is hemodynamically unstable, $TcPO_2$ will reflect changes in circulatory status.

Limits of $TcPO_2$
The values should be correlated with blood gas analysis initially to define acceptable limits for each specific situation; then $TcPO_2$ provides ideal trend monitoring. The acceptable limits for preterm babies are 53–75 mm Hg while that for term babies are 45–68 mm Hg. The values are lower for term babies because epidermis is more developed and provides increased PO_2 gradient.

Uses
Continuous monitoring of $TcPO_2$ can minimize the need for invasive blood gas sampling in NICU. It may be useful in:
a. *Weaning from ventilator*: Weaning can be faster with decreased duration of exposure to high oxygen concentrations.
b. *Decreasing hypoxia during procedures*: Endotracheal suctioning, chest physiotherapy, lumbar puncture or even tube feeding can lead to considerable hypoxic episodes which may not be given much attention in the absence of continuous monitor.
c. *Preventing hyperoxia* by continuous monitoring of $TcPO_2$ and modifying the FiO_2 appropriately, e.g. during surfactant administration.
d. *Diagnosis of right to left shunt* across *patent ductus arteriosus* by placing one sensor over right arm (pre-ductal) and over abdomen (post-ductal).
e. *Bedside PO_2 measurement* by application of blood samples to the electrode.

Other Uses
a. During surgery to avoid hypoxia and hyperoxia
b. Measurement of fetal and maternal oxygen tension during labor
c. To evaluate skin graft revascularization
d. Peripheral arterial disease mapping
e. To determine optimal level of amputation
f. During hyperbaric oxygen therapy.

Contraindications
1. Skin disorders (epidermolysis bullosa, scalded skin syndrome)

2. Relative contraindications—severe acidosis, anemia and hypotension.
3. Patients with poor skin integrity and/or adhesive allergy.

Limitations

In the presence of hypotension, poor perfusion and severe metabolic acidosis, $TcPO_2$ values reflect tissue oxygenation and not the arterial oxygenation and hence correlate poorly with PaO_2 readings. However, this discrepancy between $TcPO_2$ and PaO_2 can also be used as a diagnostic therapeutic tool to assess the peripheral tissue perfusion and see the effects of different interventions. There can be discrepancies between PaO_2 and $TcPO_2$ values if the baby is on vasoactive drugs or when the baby is having edema.

Technique

1. Familiarize yourself with the system before proceeding.
2. Perform routine electrode maintenance, if there is any question as to the status of the electrode:
    a. Remove the membrane, rinse the electrode with de-ionized water, and dry with a soft lint-free tissue or gauze.
    b. Clean the electrode using the solution provided in the cleaning kit; abrasive compounds or materials should never be used (will permanently damage the electrode).
    c. Rinse the electrode with de-ionized water and dry with lint-free tissue.
    d. Apply the electrolyte solution.
    e. Place a new membrane on the electrode. Avoid finger contact and always handle the membrane inside its protective package or with plastic tweezers.
3. Perform two-point gas calibration using the device specific apparatus, as per manufacturer's instruction.
4. Use an alcohol pad to clean and degrease the skin site where the sensor is to be placed.
5. Apply double-sided adhesive ring to the sensor.
6. Apply one drop of contact solution at the skin site.
7. Peel protective backing from adhesive ring, place sensor on the skin over the contact solution, and press the sensor to the skin.
    a. For best results, place the sensor on a location with good blood flow.
        i. Appropriate sites include the lateral abdomen, anterior or lateral chest, volar forearm, inner upper arm, inner thigh, or posterior chest.
        ii. Although large differences between pre- and post-ductal PaO_2 are uncommon, in premature infants with hyaline membrane disease, pre-ductal location of the electrode is optimal for prevention of hyperoxemia.
    b. Choose site devoid of hair.
    c. Avoid bony prominence.
    d. Avoid areas with large surface blood vessels.
8. Secure the sensor cable to prevent tugging of the electrode when the cable is manipulated.
9. Turn the site/sensor temperature control to 44°C.
10. Allow 15 to 20 minutes for site equilibration before taking readings.
11. Note the time at which the sensor was placed on the skin, so that the site can be changed after a 4-hour period (maximum site time). When changing the sensor site:
    a. Use an alcohol pad to help loosen the adhesive and peel gently from the skin.
    b. Inspect the skin site for signs of sensitivity to heat or to the adhesive. In the event of skin irritation, either lower the sensor temperature or change the site more frequently; mild erythema after sensor removal is typical.
    c. Peel adhesive ring off the sensor.
    d. Flush the membrane surface with de-ionized water.

e. Gently blot excess water and dry the sensor.
f. Recalibrate if instructed to do so by the manufacturer's guidelines.
12. Remember that response time for gas measurements is slow and values will not always immediately reflect physiologic changes.
    a. Average 90% response time for O_2 is 15 to 20 seconds.
    b. Average 90% response time for CO_2 is 60 to 90 seconds.

Newer wearable and flexible transcutaneous O_2 monitoring sensors with electrodes patterned in gas permeable membranes and electrolyte solution in non-permeable membrane which are easy to apply with less adverse effect on the skin of the babies are available.

Precautions

1. *Be aware that*
   a. Equilibration requires approximately 20 minutes once the electrode is placed, with the response time for $TcPO_2$ being much faster than that for $TcpCO_2$. Therefore, management changes based on transcutaneous values should be guided by values that have been consistent for at least 5 minutes.
   b. Periodic correlation with PO_2 from appropriate arterial sites is recommended.
   c. $TcPO_2$ may underestimate PaO_2 in the infant with hyperoxemia (PaO_2 >100 mm Hg) with reliability of $TcPO_2$ measurement decreasing as PaO_2 increases.
   d. $TcPO_2$ may underestimate PaO_2 in older infants with bronchopulmonary dysplasia.
   e. Transcutaneous blood gas measurements are affected by the state of the infant.
   f. Pressure on the sensor (e.g. infant lying on sensor) may restrict blood supply, resulting in falsely low $TcPO_2$ values.
   g. Manufactured parts are not interchangeable. Only supplies of the same brand and designated for the monitor should be used.
2. Change electrode location every 4 hours (maximum) to avoid skin burns.
3. Do not allow electrode temperature to exceed 44°C.

Common Problems

1. *Local erythema* of varying degrees depending on temperature selected for heating the electrode, duration of application and perfusion to the skin.
   Second degree *burns* occur more frequently in preterms than term babies. However, 90% of these lesions resolve by 48 hours. Hence, *the site of application should be changed every 3–4 hours.*
2. *Poor correlation of PaO_2 and $TcPO_2$*
   This could occur because of various reasons:
   a. Shock
   b. PDA with right to left shunt
   c. Temperature selection of <43°C
   d. Stab punctures (taken when baby is crying) used for comparison of PaO_2 and $TcPO_2$.

TRANSCUTANEOUS CARBON DIOXIDE ($TcPCO_2$) MONITORS

Success with $TcPCO_2$ monitoring and the fact that CO_2 diffuses through skin led to the development of $TcPCO_2$ sensors.

Principle

The $TcPCO_2$ electrode consists of a glass pH electrode, a reference electrode, heating element, electrolyte solution, teflon membrane and an amplifier. Changes in pH resulting from the reaction of diffused CO_2 with water forming carbonic acid which dissociates immediately to form H^+ and HCO^{3-} ions. The output of the electrode is inversely proportional to $PaCO_2$. The sensor is kept at 44°C and needs 2 or 1 point calibration with 5% and 10% CO_2. Like the oxygen sensor, the CO_2 sensor needs the site to be changed every 3–4 hours to avoid burns and frequent re-calibrations. The response time is 60–90 seconds.

Advantages

TcPCO$_2$ can be used in situation where continuous monitoring is indicated. It may be particularly useful when initiating and adjusting high frequency ventilation or close monitoring after extubation.

Disadvantages

It is expensive and labor intensive. Slow response time makes it less useful for detecting acute rises in PaCO$_2$. It is not useful for intermittent monitoring of a neonate with chronic lung disease.

Correlation with PaCO$_2$

Due to the metabolic contribution of local tissues, TcPCO$_2$ is typically slightly higher than PaCO$_2$. However, compared to TcPO$_2$, TcPCO$_2$ is less sensitive to changes in circulatory status. The correlation between TcPCO$_2$ and PaCO$_2$ may not be good in babies with hypoxia, shock, metabolic acidosis and chronic lung disease (CLD).

Limits of TcPCO$_2$

Like TcPO$_2$, the values should be correlated initially with PaCO$_2$; 38–60 mm Hg is generally the acceptable range. In CLD, higher values up to 75 mm Hg may be acceptable.

Monitoring

The monitoring schedule of patient and equipment during transcutaneous monitoring should be integrated into patient assessment and vital signs determinations (Table 11.1). Results should be documented in the patient's medical record and should detail the conditions under which the readings were obtained:

1. The date and time of measurement, transcutaneous reading, patient's position, respiratory rate, and activity level.
2. Inspired oxygen concentration or supplemental oxygen flow, specifying the type of oxygen delivery device.
3. Mode of ventilatory support, ventilator, or CPAP settings.
4. Electrode placement site, electrode temperature, and time of placement.
5. Results of simultaneously obtained PaO$_2$, PaCO$_2$, and pH when available.
6. Clinical appearance of patient, subjective assessment of perfusion, pallor, and skin temperature.

Recent Developments in TcPCO$_2$

TcPCO$_2$ monitors combined with SpO$_2$ monitors, reduction in the size of the probe so that it can be applied on the ear lobes of premature infants which itself increases the sensitivity of TcPCO$_2$ measurement are recent developments. Decreasing the need for recalibrating and re-membraning so that TcPCO$_2$ can be used with ease like SpO$_2$ monitors and also lowering of the sensor temperature to minimize thermal injury to the skin of the baby are other modifications.

In addition to the routine electrochemical method, optical-only detection method using near-infrared light (1580 nm) to measure the optical absorption of CO$_2$, and measurement of CO$_2$ in microoptics-type miniaturized sampling cell thereby reducing the response time to <1 min and the need for frequent calibration are also available.

TABLE 11.1: Transcutaneous monitors available in the Indian market

S.No.	Make	Principles	Dealers	Unit cost (₹)
1.	TCM3 (TcPO$_2$ + TcPCO$_2$)	Radiometer	SBP medicare	2,00,000
2.	Microgas 7640	Kontron	L and T	2,00,000
3.	Novametrix	Novametrix	Rustagi Surgicals	4,00,000

Frequently Asked Questions (FAQs)

Q. 1. In which clinical situations transcutaneous monitoring may be useful?
ELBW neonates (<1000 gm) with thin skin during the first 2 to 3 weeks of life, in whom accurate estimation of PaO_2 and $PaCO_2$ is desired continuously and for prolonged time. Its use will decrease the frequency of arterial blood gas analysis.

Q. 2. What precautions need to be kept in mind while using it?
Do not forget to change the site of application of the electrode every 3 to 4 hrs. Calibration of electrode needs to be done periodically. Transcutaneous blood gas monitoring should be continuous for development of trending data. So-called spot checks are not appropriate. Take into account the site of transcutaneous sensor (pre-ductal *vs* post-ductal) and the site of arterial line when comparing PaO_2 values. In hemodynamically unstable patients the transcutaneous PaO_2 and $PaCO_2$ values are influenced by circulatory status.

Q. 3. How will the readings of the transcutaneous monitor help me?
The displayed values represent more or less the blood gases. The monitor values need to be correlated with the blood gas values, then they provide useful trending pattern for a particular patient. This will reduce the need of frequent blood gas sampling. A rising transcutaneous pCO_2 should always be considered clinically significant until proved otherwise: It may indicate decreasing peripheral perfusion. It can also alert staff to a pneumothorax before other clinical signs.

Q. 4. When should I buy this instrument?
If you are using oxygen saturation monitors for ventilated sick babies and you have blood gas analyzer available with you, transcutaneous monitor will be a useful adjunct.

Q. 5. How does transcutaneous oxygen monitor compare with a pulse oximeter?
See table below.

Q. 6. How does transcutaneous carbon dioxide compare with arterial CO_2?
Reported correlation are 0.90 to 0.93 agreeing with $PaCO_2 \pm 4$ mm Hg. Sensors require 15 to 20 minutes of equilibration before providing stable readings. Due to contribution of CO_2 produced and added to the arterial CO_2 by local tissue, the values are slightly higher.

Q. 7. Do $TcPCO_2$ monitors need calibration and special maintenance?
Before use, the $TcPCO_2$ sensor must be calibrated against a standard CO_2 gas mixture, as they tend to drift during use. Electrodes are costly and fragile, and replacement of membranes is needed periodically. Sensor sites must be rotated every 3 to 4 hours to avoid burns.

	Transcutaneous monitor	*Pulse oximeter*
Calibration	8–10 min every 4 hours	None (1–3 min with current models)
Warm-up time	20 min after probe application- for capillary bed to arterialize	None
Lag time	20–90 sec	None
Complications	Burns	Pressure damage to skin
Artifacts	Membrane wrinkles Air between membrane and skin Pressure on probe	Movement of extremity Ambient light
CO_2 status	Measured by $TcPCO_2$ electrode	Not measured
Detection of hyperoxia	Reliable	Unreliable

Q. 8. What are the clinical situations where there is poor correlation of $TcPO_2$?

Presence of shock, use of high dose tolazoline, isoprenaline, dopamine; obstructive heart disease with hypoperfusion, edema, severe hypothermia will lead to $TcPO_2 < PaO_2$.

Right to left ductal shunt with pre-ductal electrode and post-ductal arterial sample will result in $TcPO_2 > PaO_2$.

Q. 9. What are the technical reasons for poor correlation of $TcPO_2$ and PaO_2?

Poor correlation of $TcPO_2$ and PaO_2 (see table below).

Q. 10. Do transcutaneous monitors need validation against some gold standard?

Arterial blood gas values should be compared to transcutaneous readings taken at the time of arterial sampling in order to validate the transcutaneous values. This validation should be performed initially and periodically as dictated by clinical condition of the patient.

Q. 11. Is there any single technique which measures both the $TcPO_2$ and $TcPCO_2$ simultaneously?

Mass spectrometry. *Principle*: In this technique a stainless steel, self-heated, flat surface probe covered with a Mylar membrane admits gases to a mass spectrometer for direct measurement of PO_2 and pCO_2.

	Technical reason	
	Problem	*Solution*
$TcPO_2 < PaO_2$	1. Improper calibration	Recalibrate
	2. Insufficient warm up period after electrode application	Allow longer warm-up period
	3. Insufficient heating temperature	Increase heating temperature
$TcPO_2 > PaO_2$	1. Improper calibration	Recalibrate
	2. $TcpO_2$ reading taken immediately after electrode application	Allow longer warm-period
	3. Air bubble beneath membrane or leak to atmosphere	Reapply electrode
	4. Excessive heating temperature	Attempt calibration at lower temperature

Chapter 12

Capnograph (End-Tidal Carbon Dioxide Monitors)

The displayed or recorded waveform of carbon dioxide content in the respired gases is called capnography. The instrument used to record the partial pressures of CO_2 in expired air is called capnograph. It provides a noninvasive means of respiratory monitoring. Capnography is useful for much more than checking the position of the endotracheal tube. It provides information about CO_2 production, pulmonary perfusion, alveolar ventilation, respiratory patterns and elimination of CO_2. Thus, it gives us a rapid and reliable method to detect life-threatening conditions such as malposition of tracheal tubes, ventilatory failure, circulatory failure and defective breathing circuits. The technique depends on the achievement of an end-tidal CO_2 ($EtCO_2$) plateau from which alveolar CO_2 is estimated. It is, therefore, rate and flow dependent. In newborns, the relatively high respiratory rates and low tidal volumes and marked ventilation-perfusion mismatching mean that a stable $EtCO_2$ is often not achieved. This results in inconsistent underestimation of $PaCO_2$. Correlation coefficients with $PaCO_2$ vary from 0.69 to 0.92. Newer systems with lower dead space and mainstream sensors may make this method more accurate in future.

Two primary methods are available to monitor CO_2 gas

a. *Mass spectrometer*: This is usually a centralized monitoring system serving more than one patient, possibly as many as 20. The major disadvantages of mass spectrometry are long response time in neonates and obstruction of sampling tubes with mucus.

b. *Infrared detector*: They are used to monitor individual patient and function as stand-alone systems. It works on the principle that molecules containing more than one element absorb infrared light in a unique and characteristic manner. In addition, every substance absorbs light at a characteristic wavelength. CO_2 absorbs infrared light at 2600 nm and 4300 nm wavelengths. Infrared capnograph use a hot wire to emit infrared light and a filter to obtain the desired wavelength. The infrared light is beamed through a calibrated CO_2 filled chamber serving as a control and through the chamber with the gas to be analyzed. Infrared sensitive photocells receive light from both chambers and calculate the concentration of CO_2 in the sample gas by comparing it with the known concentrations of CO_2 in the control. The amount of light absorbed is dependent on the concentrations of CO_2 molecules in the sample. A semiconductor called the detector is used to create an electrical signal that can be processed to display continuous CO_2 concentration. The response time is approximately 100 msec.

Infrared devices are subdivided into two categories

i. *The mainstream analyzer*: It has a special flow through an adapter, called the couvette, which is mounted directly in-line with the endotracheal tube and ventilator circuit. It contains an infrared light source and photodetector.

ii. *The side-stream analyzer*: It uses suction to withdraw a continuous sample of inspired and expired gas through a capillary tube, from the patient's airway to the monitor. Narrow lumen of the sampling tube makes it prone to be blocked by pulmonary secretions.

Physiological Considerations

When expiratory CO_2 concentration is measured for the purpose of assessing adequacy of alveolar ventilation, it is intended to reflect CO_2 in arterial blood. In a normal lung, CO_2 rapidly diffuses across the capillary alveolar membrane when ventilation and perfusion are well matched. End-tidal CO_2 is the partial pressure equivalent of airway CO_2 concentration at the end of a tidal breath. Since alveolar CO_2 ($PaCO_2$) approximates $PaCO_2$, it is theoretically possible to obtain an accurate estimate of $PaCO_2$ from measurement of the pCO_2 of exhaled gas. However, the concentration of CO_2 measured during exhalation is greatly influenced by various physiologic components, which include ventilation–perfusion ratios within the lungs, total CO_2 production, and total alveolar ventilation. In critical situations, these variables may not be stable. Measurement of peak expired CO_2 underestimates $PaCO_2$ as a result of these physiologic factors. An algorithm using a first measure difference between $PaCO_2$ and peak expired CO_2 in a given infant as an additive factor to a new measurement of peak expired CO_2 results in an accurate estimate within a reasonable limit. Efforts have been made to analyze the influence of the relationship between ventilation–perfusion and in the correlation between $PaCO_2$ and end-tidal CO_2. It was observed that end-tidal CO_2 measurement was an effective and accurate technique for the monitoring of newborns when a/A O_2 ratio were >0.3.

Normal Capnogram (Fig. 12.1)

At the end of normal inhalation, the concentration of CO_2 is highest in the alveoli; the concentration gradually decreases proximally in the respiratory tree. At the beginning of exhalation, the first gas sampled is the CO_2 – free from tracheal dead space. The capnogram displays a segment corresponding to zero CO_2 concentration. As the exhalation continues, detection of CO_2 is displayed in segment A–B. Slowly changing CO_2 concentration over time produces a nearly horizontal alveolar plateau segment B–C. Point C represents end-tidal CO_2 that is the best approximation of arterial

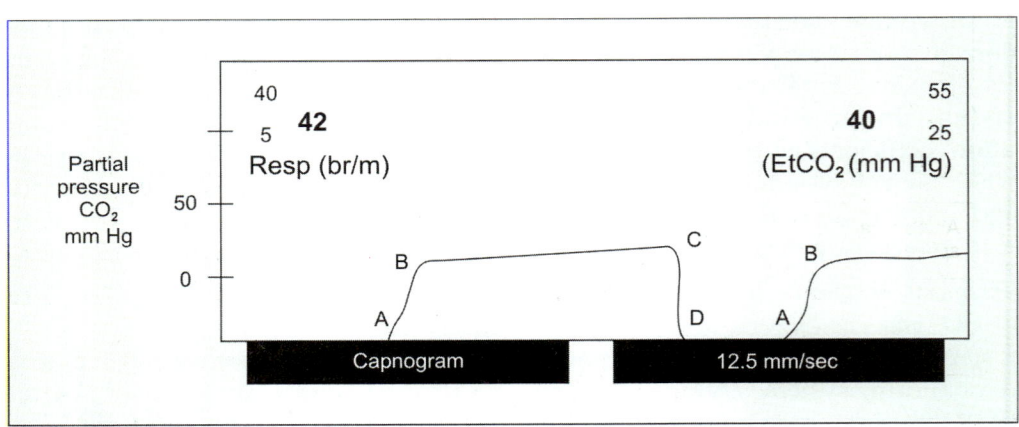

Fig. 12.1: A typical end-tidal CO_2 monitor wave

pCO_2. Inhalation that is normally free of CO_2 gas causes the tracing to return to the zero base line. There are three types of capnographs: trend, time and volume capnograph. The following figure displays a time capnograph.

Advantages

$EtCO_2$ may provide intermittent trend monitoring for larger intubated infants with chronic lung disease. It can be used to detect airflow obstruction and apnea in non-intubated infants by measuring at the nostril level.

Disadvantages

The additional dead space introduced by the airway adapter can cause CO_2 retention as much as 6 to 10 mm of Hg. It cannot be employed during high frequency ventilation.

Clinical Applications of Capnography

Capnographic monitoring provides a non-invasive means to evaluate the integrity of the airway as well as breathing circuit to the quality of a patient's cardiopulmonary function, and malfunctions can often be detected by changes in the capnogram.

a. *Sudden drop of end-tidal CO_2 to zero* or near zero value occurs in esophageal intubation, complete airway disconnection, total obstruction of endotracheal tube, complete ventilator malfunction.
   All these events are potentially fatal airway disasters and require immediate examination of the patient and the ventilator circuit. Capnography would greatly enhance detection of such a catastrophic event more quickly and accurately in comparison with clinical indicators.

b. *Progressive drop in end-tidal CO_2* occurring over time signals, a potentially serious event involving the cardiopulmonary system, such as sudden hypotension due to massive blood loss, circulatory arrest, pulmonary embolism due to thrombus or air, etc.

c. *Fall in end-tidal CO_2 approaching with loss of plateau* denotes absence of full exhalations. It occurs in loosely fitting endotracheal tube, partial obstruction of the endotracheal tube, and partial ventilator disconnection.

d. *Wide $PaCO_2-EtCO_2$ gradient* represents excessive dead space ventilation. It may occur in pneumonia, bronchopulmonary dysplasia and hyaline membrane disease (Fig. 12.2).

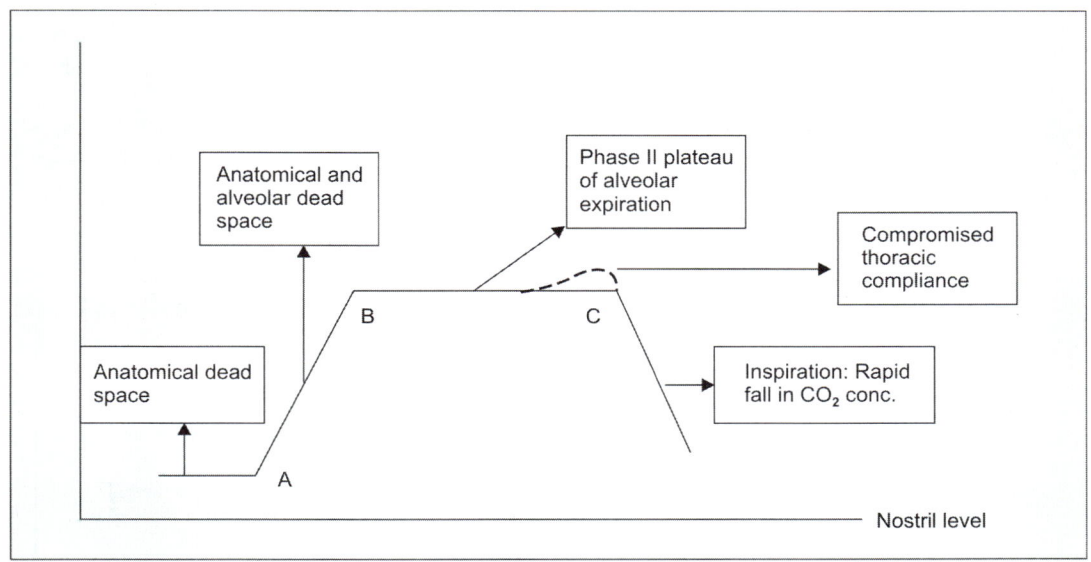

Fig. 12.2: Normal waveform of $EtCO_2$

e. *Progressive drop in end-tidal CO_2 with normal plateau* may occur in falling body temperature, slowly decreasing systemic or pulmonary perfusion, hyperventilation, and reduced CO_2 production.
f. *Steady rise in end-tidal CO_2 without morphologic changes* denotes partial airway obstruction, rising body temperature, leak in the ventilator circuit with hypoventilation.
g. *Acute and transient increase in end-tidal CO_2* may occur in sodium bicarbonate administration acute increase in cardiac output.
h. *Gradual rise in baseline and end-tidal CO_2* occurs when previously exhaled CO_2 is being re-breathed from the circuit. The inspiratory portion of the capnogram fails to reach the zero baselines.
i. *End-tidal CO_2 monitoring during conscious sedation for painful procedures* detects subclinical respiratory depression earlier than pulse oximetry and respiratory rate alone.
j. *End-tidal CO_2 monitoring during transport* may prevent hypocarbia caused by unintentional hyperventilation. This is a vital issue in regulation of cerebral blood flow.
k. *Continuous end-tidal CO_2 monitoring* may help avoid complications of hypocarbia and hypercarbia. The technical advances and the introduction of surfactant therapy, which improves ventilation–perfusion matching, might improve the clinical utility of end-tidal monitoring (Table 12.1).

Conclusion

Capnography has not yet achieved wide application as a noninvasive method for monitoring and optimizing assisted ventilation in neonatal intensive care units. However, available literature reveals that capnography may be an effective monitoring tool to rapidly identify life-threatening situations in a critically sick neonate. Nevertheless, it is important to note that capnography is a qualitative rather than quantitative technique. It detects the abnormal events without measuring the degree of physiologic derangement.

TABLE 12.1: Common brands of end-tidal carbon dioxide monitors in the market

S.No.	Make	Principals	Dealers	Unit cost (₹)
1.	Capnograph	Novometrix NPB 70 NPB 75 CO_2 SMO	Rustagi Surgicals Systems Biomedical	2,00,000-300,000 3,00,000/-
2.	Oxicap 4700 Ohmeda 5200 or (measures inspired CO_2 also)	Datex-Ohmeda	Datex Ohmeda Phoenix, Medisphere	Cost of $EtCO_2$ module installation 2,50,000 to 300,000
3.	Datascope	BCI	—	2,50,000/-
4.	Criticare 602 series	Criticare	Criticare Systems India	3,50,000/-
5.	Normocap Oxy	Datex	AVL Biomedicals	2,50,000/-
6.	Minipack 300C	Pacetech	Medex	2,50,000/-

Frequently Asked Questions (FAQs)

Q. 1. How should one interpret the reading of an end-tidal CO_2 monitor?

End-tidal CO_2 monitor reflects more or less arterial pCO_2. Though the values correlate fairly well for pCO_2 ranges of 40–60 mm of Hg, they are not reliable for low or high values. It may not be a bad idea that the values are initially correlated to arterial blood gas pCO_2 values and later on, the trend followed. This will curtail the need for doing frequent blood gas sampling.

Q. 2. Enumerate conditions where capnography is useful.

i. Verification of ET position
ii. CO_2 elimination during cardiac arrest or CPR
iii. Hypo- or hyperventilation
iv. Rebreathing of CO_2
v. Obstructed airway
vi. Inadequate seal of ET.

Q. 3. What clinical decision-making processes are facilitated by use of capnography?

This helps in
i. Customizing the respiratory support to a baby on ventilator.
ii. Getting an idea about the changing clinical condition of ventilated baby.
iii. Taking a decision about the weaning.
iv. Getting an idea about the extent of lung disease.

Q. 4. How do I calibrate my monitor?

The machine is first zeroed to room air and supplied with a pure CO_2 cylinder 5% or in a specified concentration as specified. It should be calibrated every 8 to 12 hrs with standard CO_2 gas mixture.

Q. 5. In addition to making possible judgement of clinical situation based on wave pattern, what are the advantages of end-tidal CO_2?

End-tidal CO_2 monitoring will reduce the need for frequent blood sampling for arterial blood gas. The trends generated may guide the ventilator settings in a clinical situation.

Q. 6. Kindly give examples and significance of typical capnography waveforms?

Examples of typical capnogram waveforms are:

Normal waveform: The normal capnogram provides a waveform of changing levels of expired CO_2 (refer to Fig. 12.2)
- A–B: Ascending limb, expiratory portion, mixed dead space and alveolar air with increasing concentration of CO_2. Normally starts at a baseline of zero.
- B–C: Alveolar plateau contains mixed alveolar gas with end-tidal measured at C.
- C–D: Descending limb, inspiratory portion, with decreasing concentration of CO_2. Normally returns to a baseline of zero.

Cardiogenic oscillations: Cardiogenic oscillations appear during the final phase of the alveolar plateau and during the descending limb. They are caused by the heart-beating against the lungs (Fig. 12.3).

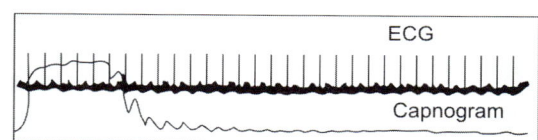

Fig. 12.3: Cardiogenic oscillations in capnograph

Characteristics
- Rhythmic and equal to heart rate.
- May be observed in pediatric patients mechanically ventilated at low respiratory rates with prolonged expiratory times.
- Ripples during phase II and phase III occur due to changes in pulmonary blood volume and ultimately CO_2 pressure as a result of cardiac contractions.

Hypoventilation: An increase in the level of the end-tidal CO_2 from previous levels, steady increase in phase II without changing the baseline (Fig. 12.4).

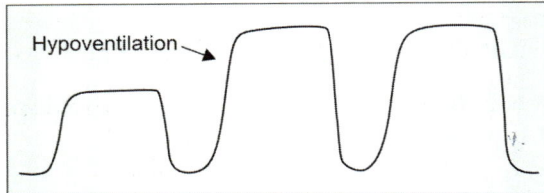

Fig. 12.4: Waveforms in hypoventilation

Possible causes
- Decrease in respiratory rate
- Decrease in tidal volume
- Increase in metabolic rate
- Rapid rise in body temperature (malignant hyperthermia).

Hyperventilation: A decrease in the level of the end-tidal CO_2 from previous levels, steady decrease in phase II without changing the baseline (Fig. 12.5).

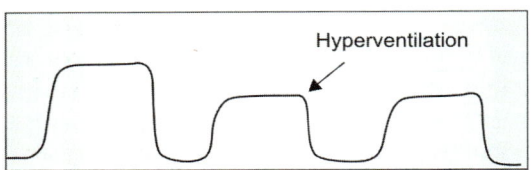

Fig. 12.5: Waveforms in hyperventilation

Possible causes
- Increase in respiratory rate
- Increase in tidal volume
- Decrease in metabolic rate
- Fall in body temperature

Muscle relaxants: Clefts are seen in the final third portion of the alveolar plateau. They appear when the action of the muscle relaxants are affected by spontaneous ventilation (Fig. 12.6).

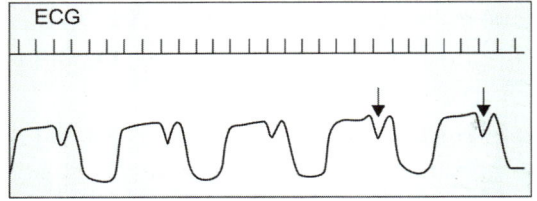

Fig. 12.6: Waveforms after use of muscle relaxants

Characteristics
- Depth of the cleft is inversely proportional to the degree of drug activity.
- Position fairly constant on same patient but may not be present in every capnogram.

Re-breathing: Re-breathing is characterized by an elevation in the baseline, with elevation in phase II and decreasing inspiratory efforts, with a corresponding increase in end-tidal CO_2. It indicates the re-breathing of the previously exhaled CO_2 (Fig. 12.7).

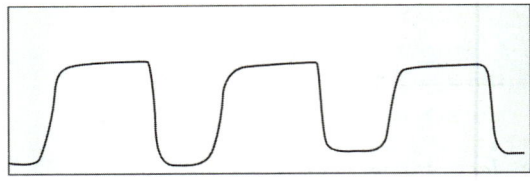

Fig. 12.7: Waveforms in re-breathing

Possible causes
- Insufficient expiratory time
- Faulty expiratory valve
- Inadequate inspiratory flow
- Malfunction of a CO_2 absorber system
- Partial re-breathing circuits

Obstruction in breathing circuit or airway: An obstruction to the expiratory gas flow noted as a change in the slope of the ascending limb of the capnogram. The expiratory portion may diminish without a plateau (Fig. 12.8).

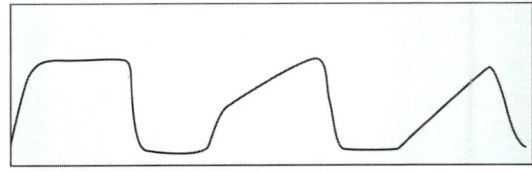

Fig. 12.8: Waveforms due to obstruction in airway

Possible causes
- Partial obstruction in the expiratory limb of the breathing circuit.
- Presence of a foreign body in the upper airway
- Partially kinked or occluded artificial airway

- Herniated endotracheal/tracheostomy tube cuff
- Bronchospasm.

Endotracheal tube kinked: Any obstruction will cause an abrupt change in the ascending limb resulting in either a diminished plateau or no plateau. End-tidal CO_2 and slope will depend on the degree of obstruction (Fig. 12.9).

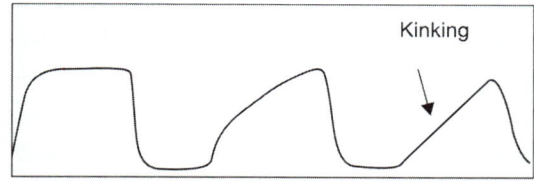

Fig. 12.9: Waveforms caused by kinked ET tube

Inadequate seal leak around endotracheal tube: A capnogram in which the downward slope of the plateau blends in with the descending limb (Fig. 12.10).

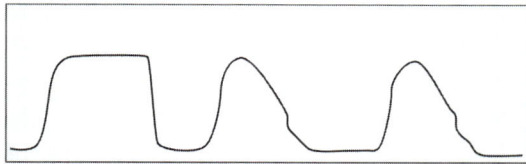

Fig. 12.10: Waveforms due to ET tube leak

Possible causes
- A leaky or deflated endotracheal or tracheostomy cuff
- An artificial airway that is too small for the patient.

Endotracheal tube in esophagus: A normal capnogram is the best available evidence that the ET tube is correctly positioned and that ventilation is occurring. When the ET tube is placed in the esophagus, either no CO_2 is sensed or only small transient capnograms are present (Fig. 12.11).

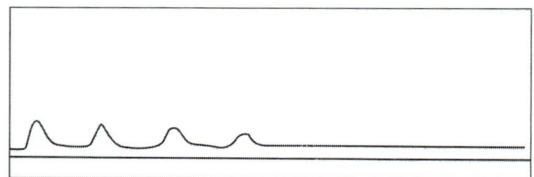

Fig. 12.11: Waveforms in esophageal intubation

Faulty ventilator circuit valve: Waveform evaluation
- Baseline elevated
- Sloping descending limb of capnogram
- Allows patient to re-breathe exhaled gas.

Spontaneous breathing: Short alveolar plateau, with increased frequency.

Chapter 13

Ultrasound Machine

Ultrasonography is a useful modality in neonatal intensive care unit. Bedside ultrasound provides safe and an effective means to screen neuroaxis of sick newborn in the thermoprotected nursery environment. Ultrasound machine basically consists of a computer and a transducer. The transducer picks up the ultrasound waves generated by transducer which travels and get refracted by the underlying tissues. The computer analyzes the waves and produces two-dimensional image on the screen.

Principle

Ultrasound is the name given to high frequency sound waves beyond the normal audible range (20 to 20000 Hz). The usual range lies between 2 to 10 MHz. The nature of image produced by the refracted waves depends on composition of the underlying tissues thus projecting differential image. Soft tissue would produce blacker image (echo lucent), air relatively white, fluid as black, while underlying bone would appear as complete white image (echo dense). The positional information of an organ is obtained by time taken by echo to return to transducer. If echo-generating structure is moving then echo will have different frequency than that of pulse emitted from transducer. This is known as Doppler effect, which is utilized for blood flow detection. Higher frequency transducer would have better resolution but the depth of penetration in the tissue would be lesser.

DISPLAY MODES

This refers to different application of the ultrasound waves:

1. *A mode (amplitude modulation)*: Produces a one-dimensional image displaying the returning echo signals in the form of amplitudes along the vertical axis and time along the horizontal axis. The greater the reflection at the tissue level taller the amplitude spike will be. It is mainly used for ophthalmic purposes.

2. *B mode (brightness modulation)*: Displays the intensity of the returning echo by varying the brightness of a dot so that the different tissues appear in different level of brightness depending upon the composition. This is the most commonly used mode and is the basis for all real time imaging.

3. *M mode (motion mode)*: Displays time along the horizontal axis and depth along the vertical axis to depict the movements, especially in cardiac structures. Its main use is for taking different measurement in cardiac sonography.

Ultrasonography is a real time imaging. It provides a dynamic presentation of multiple image frames per second over selected areas of body. The frame rate is dependent on the frequency and depth.

Doppler principle refers to a change in frequency when the motion of laminar or turbulent flow is detected within a vascular structure. The common uses in NICU are to detect flow across the PDA so as to define its hemodynamic significance and to detect catheter-associated thrombosis.

ULTRASOUND OF BRAIN

It requires multiple views:

1. *Sagittal scan*: The transducer is centered at anterior fontanel with scanning axis in sagittal plane (Fig. 13.1). Angle the transducer first towards right and then left to see the different structures. Midline structure shows the corpus callosum, third, fourth ventricles and posterior fossa. Parasagittal sections show lateral ventricles while extreme parasagittal section shows insula.

2. *Coronal section*: The transducer is rotated at 90° so that scan plane is aligned transversely. The beam is angled forward and backward. From before backward, it scans through frontal lobes, frontal horns, foramen of Monroe, third ventricle, trigone of ventricles at the level of choroid plexus and the occipital horns (Fig. 13.2).

3. *Axial scans*: The transducer should be centered just above the ears at lateral fontanels. Angle the beam up towards the

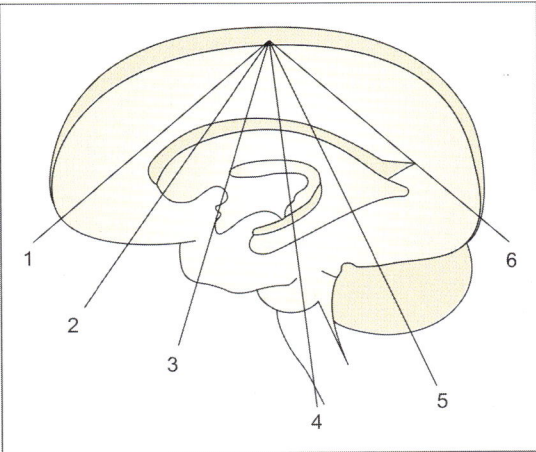

Fig.13.2: Six standard coronal sections

vault and then down towards the base of skull on both sides. It shows from below upwards pedicles, pulsation of circle of Willis, thalamus, falx cerebri and walls of the lateral ventricles.

TRANSDUCER

Ultrasound transducers often also known as probes or scan heads form backbone of ultrasound machine. Ultrasound probe acts on principle of piezoelectricity, which was discovered in 1880. As per this principle, voltage applied to certain type of materials results in production of pressure waves. A transducer converts one type of energy into another. Based on "Pulse-Echo" principle the transducer crystals converts:

- Electricity into sound = 'Pulse'
- Sound into electricity = 'Echo'

Various formulations of lead zerconate titanate (PZT) are commonly used material for construction of the transducer system, which is sometimes, referred as piezoelectric crystal.

Previously mechanical transducers were in use, which used to be constructed using single crystal system. Now-a-days these have been replaced by electronic transducers, which contain multiple vibration-producing units. The transducers are mainly of three types:

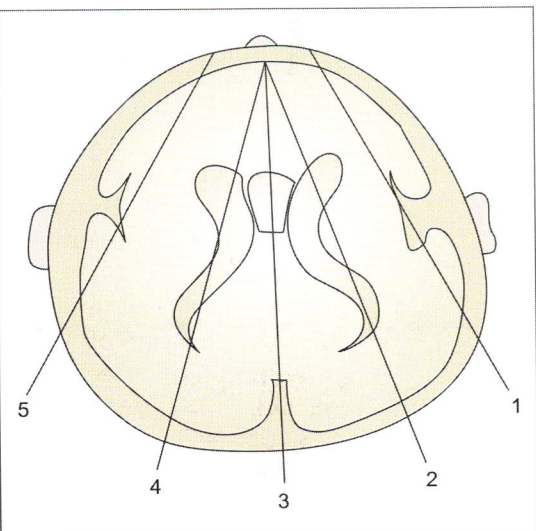

Fig. 13.1: Five standard sagittal planes

1. ***Linear array transducers:*** These probes are rectangular in shape. Here each pulse of sound wave starts from different points from the transducer and travel in parallel to each other. The resultant image is rectangular. These transducers are used for abdominal or obstetrical purposes.
2. ***Sector transducer:*** Sometimes also known as microconvex transducers. Here each pulse of sound wave originate from same starting point but go out in different direction. The resultant scan on screen is triangular or fan shaped. These types of transducers need a very small scanning window to peep inside the organ system.
3. ***Convex transducer:*** It produces scans that are somewhere between linear and sector probes. It is useful for the entire body except for echocardiography.

The selection of appropriate transducer depends on the intended uses, e.g. for abdominal or cardiac purpose. The transducers can be given a desired shape for example for using inside a body cavity or to fit on a finger for its intraoperative use. *For neonatal purposes, a sector transducer of 5–7.5 MHz is ideal especially for scanning brain. It can be used for abdominal purposes also, in neonates* (Fig. 13.3 and Table 13.1).

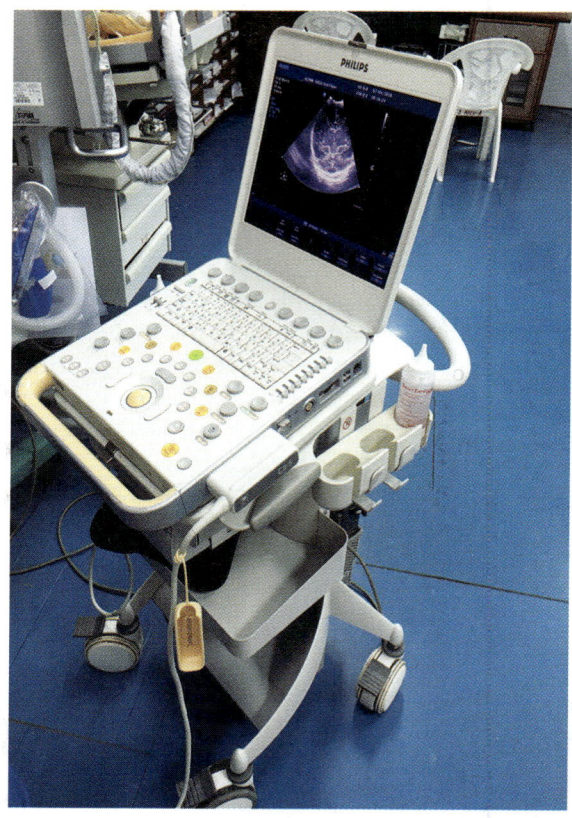

Fig. 13.3: Advanced ultrasound machine with facility of Echo and colored Doppler

TABLE 13.1: Common brands of ultrasound machine available in the market

S.No.	Make	Features	Principals	Approximate cost (₹)*
1.	Sonosite M180 with 4–7 MHz broad band curved array probe	• Extremely compact (33.8 cm × 19.3 cm × 6.1 cm) and light weight (2.4 kg)	Philips Healthcare	10–13 lakhs
2.	Scanner 100S with mechanical probe	• Battery backup • Image storage on floppy • 9 inch display screen	Philips Healthcare	5 lakhs
3.	a. Sonoline-Adara b. Acuson X 300	• 12 inch display screen • Storage on 3.5 inch floppy	Siemens	7.5 lakhs 20 lakhs
4.	Clarify	• Upgradeable to B/W Doppler • Memory for 9 images • 9 inch display screen	Larsen and Toubero Ltd.	7.5 lakhs

(Contd...)

TABLE 13.1: Common brands of ultrasound machine available in the market (*Contd...*)

S.No.	Make	Features	Principals	Approximate cost (₹)*
5.	a. Justvision 200 multiple frequency selection probe (5/6/7 MHz probe) b. NexMio SG SSA 580A	• Compact and mobile • Compact and mobile • 9 inch display screen	Toshiba	a. 6 lakhs b. 21 Lakhs
6.	a. SSD 900 convex sector/linear scanner b. SSD 4000 Terason t 3000	• 9 inch display screen • 9 inch display screen	Aloka Co. Ltd.	a. 7 lakhs b. 15 lakhs
7.	Logiq E 200	15 inch display screen	Wipro GE Healthcare	14 lakhs
8.	Micromax	Display: 10.4"/26.4 cm diagonal LCD	Seonie Technocrats Sonosite Inc	15 lakhs
9.	Epiq7	12" touch screen	Philips Healthcare	16 lakhs
10.	Vivid E9	4D cardiac imaging and ergonomics based	Wipro GE	20 lakhs

*Cost varies based on number/type of probes, facility for echo. High end machines may cost 30–35 lakhs

Common Indications in Neonates

Ultrasound should be the first investigation in the following situations:

1. ***Intracranial hemorrhage:*** Ultrasound is an excellent modality to diagnose and to follow subependymal—intraventricular, intraparenchymal hemorrhages. There are various grading described based on ultrasound findings. Severe grades of bleeding can result in hydrocephalus or porencephalic cyst.
2. ***Hydrocephalus:*** Normally the ratio of the ventricular diameter to the hemispheric diameter is less than 1:3. If it is more, hydrocephalus must be considered.
3. ***Periventricular leukomalacia:*** Especially in later stages after 2–3 weeks can be picked up very well.
4. ***Cerebral edema:*** Leads to obliteration of ventricles and sulcus. The brain is more echogenic than usual.
5. Cerebral calcifications and cysts.
6. ***Abdominal and pelvic organs:*** Evaluation of acute renal failure to rule out renal dysgenesis, aplasia, any obstructive uropathy to facilitate the management.
7. ***Echocardiographic examinations*** can be performed in presence of Doppler facility.
8. ***Ultrasound-guided procedures*** such as thoracentesis, ureteronephrostomy, and suprapubic puncture of bladder.

Advantages

1. The procedure is carried out in the thermo-protected environment without disturbing the baby with all its equipment attached. There is no need to transport the baby, which can be dangerous when the baby is sick.
2. There is no radiation hazard. It is safe.
3. It has valuable diagnostic and prognostic value.

Limitations

1. It is operator dependent so one needs to train himself/herself. It is an easy task to learn to screen for common conditions.
2. Less sensitive for detection of subarachnoid hemorrhage, posterior fossa lesions, cerebral edema.

3. Early stages of periventricular leukomalacia, intraventricular hemorrhage or early changes of HIE can be missed.

Maintenance

1. Protect the equipment from dust or extremes of temperature (air-conditioning is not essential).
2. Prevent the transducers from fall as it can lead to malfunctioning of the equipment.
3. Clean the jelly applied on the transducer.
4. Run on stabilized electrical voltage.

Frequently Asked Questions (FAQs)

Q. 1. When should one buy an ultrasound machine in a level II neonatal unit?
Ultrasound machine certainly should not be purchased at the beginning. One should invest in equipment needed for basic neonatal care such as open care system, incubators, phototherapy units, etc. In a later phase when you have a level III neonatal care set with reasonable survival of small babies below 1500 gm then you can think of acquiring ultrasound machine. It adds to quality improvement of neonatal care and helps you make appropriate decision in critical situations.

Q. 2. What specification of machine would meet the needs of NICU?
There are varieties of ultrasound machines available in the market with different features (e.g. with Doppler facility). Primarily it should match your budget requirement and felt needs. In addition, it should be a mobile and portable unit, compact, with good resolution and user friendly. Documentation facility is desirable.

Q. 3. What are the salient points for maintenance of the machine?
It should be guided by manufacturer's recommendation. Try to protect from dust as the computer chip may be damaged. The transducer should be carefully kept and should be prevented from fall. The jelly is a good food for rats and they will eat it with transducer. Clean it properly after use.

Q. 4. Who should scan the babies—neonatologist or radiologist?
Normally pediatricians are not trained in ultrasound diagnosis but one may not have radiologist available at bedside at odd times for screening. The best option is to screen yourself and then get it confirmed by radiologist. It is not at all difficult to acquire the skill with a little practice. Even radiologists are not comfortable with neonatal scan, so the best is pediatrician anyway *(believe it!)*.

Q. 5. How to prevent cross infection with use of a single probe on different babies?
The probe should be cleaned before and after use on the baby. Agents that can be used for cleaning include quarternary ammonium, mild soap and water or chlorhexidine solution. The ultrasound machine should also be cleaned from outside by detergent solution using soft cloth. One should observe routine asepsis measures while scanning a neonate.

Chapter 14

Cerebral Function Monitor

The cerebral function monitor (CFM), or amplitude-integrated EEG (aEEG), is a device for monitoring background neurological activity. CFM was first developed in the late 1960s to monitor adults undergoing surgery, suffering head trauma, or in coma. In the mid 1980s it was first used in Europe to monitor neonates. Since then a number of researchers has demonstrated its utility in monitoring neonates with perinatal asphyxia. It is also shown to be useful in detection of seizures in the neonates. It has the potential to become a popular tool in cerebral monitoring of the neonates, primarily because of its simplicity and ease of interpretation.

ACQUIRING DATA ON A CFM

The CFM (cerebral function monitor) records a single channel of EEG. For this, it uses a single, biparietal or frontal lead (three electrodes) to obtain an EEG signal. The acquired signal is filtered, rectified, and semilogarithmically compressed. Frequencies <2 and >15 Hz are selectively filtered to reduce artifacts caused by movement, ECG and other electronic equipment. The trace is displayed on the monitor and can be printed on a graph paper using a thermal printer similar to an ECG machine. Some machines simultaneously also display the conventional EEG record picked up from the electrodes.

PLACEMENT OF ELECTRODES

Usually the signal is recorded from two electrodes placed on either side of the head. A third electrode acts as a ground electrode (Fig. 14.1). The optimal location is biparietal (P3/P4), as this is the area to be least affected by scalp muscle activity and eye movement artifacts.

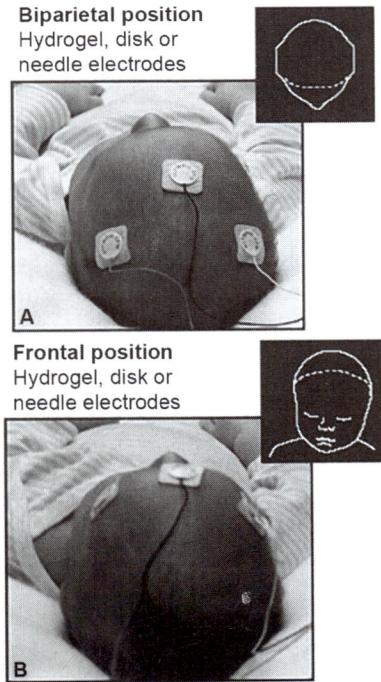

Fig. 14.1: (A) In the preferred method there is a central electrode at the vertex and two electrodes in the parietal region; (B) Alternative frontal placement of all the three electrodes

Type of Electrodes

There are *two* types of recording electrodes:
1. ***Cup electrodes***: These require meticulous preparation before application.
2. ***Low impedance needle electrodes***: Small needle electrodes placed subdermally and secured with a tape. They have low impedance, are not dislodged easily and can be retained for several days.

Differences from an EEG Recording

The acquired signal undergoes filtering, rectification and compression. The output is displayed at a very slow chart speed, 1 mm per minute, giving a trace in which neither the configuration nor the amplitude of individual brain waves can be identified. Due to the processing, the output is no longer a regular EEG signal but is, rather, a representation of the overall electrocortical background activity of the brain.

Interpretation of aEEG

The CFM record consists of a dense trace that can vary in width. Two features of the trace are assessed:
1. Amplitude
2. Seizure activity

1. ***Amplitude***: It is assessed by measuring the upper and lower margins of the trace against the scale printed on the printer paper. Normally the CFM appears as a trace varying from about 10–40 microvolts. Based on amplitude the following information is obtained:
   - Background amplitude
   - The presence and degree of discontinuity
   - Presence of burst suppression.
2. ***Seizures***: Apart from giving an idea of the background activity, aEEG tracing is helpful to identify seizures. They appear as sudden rise and narrowing of the trace.

Advantages

1. Allows continuous, real time prolonged monitoring of cerebral activity.
2. Allows identification of seizures
3. Easy to use and interpret
4. Potentially enhances newborn care by allowing early decisions on initiating and tailoring therapy.

Disadvantages

1. It does not give information about EEG frequency.
2. Because of the few number of electrodes used, it cannot be used to detect focal abnormalities.
3. Can miss seizure activity of shorter duration.

RECOGNIZABLE PATTERNS OF AEEG ON A CFM

1. Normal Trace

In normal term neonates the width of the trace varies from approximately 5–25 (max 50) microvolts (Fig. 14.2). The width of the trace also varies with arousal, medication exposure and gestational age of the neonate. The trace is narrower when the neonate is awake and widens during sleep. These changes in the width of the trace with the physiological state are called sleep/wake cycling.

2. Moderately Abnormal Trace

A moderately abnormal trace has an upper margin that is greater than 10 microvolts and a lower margin less than 5 microvolts. The trace therefore appears wider than in normal infants. This appearance can be seen in following situations:
- Neonates with moderately severe encephalopathy.
- Immediately after administration of drugs such as anticonvulsants and sedatives.
- Preterm infants (below 37 weeks gestation).

3. Severely Abnormal Trace

A severely abnormal trace is characterized by a general suppression of amplitude so that the trace appears narrow and of low voltage. The

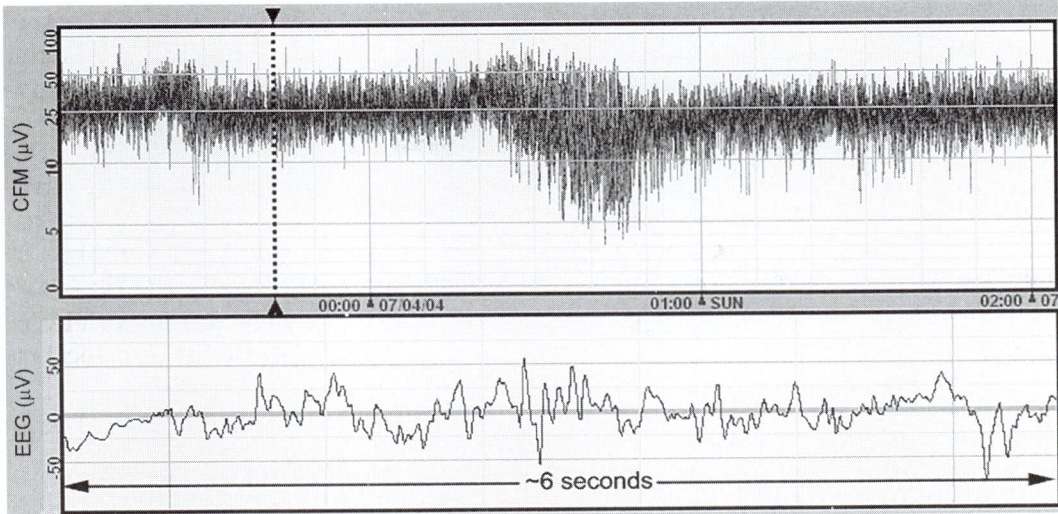

Fig.14.2: Normal trace with EEG

upper margin of the trace is less than 10 microvolts. The lower margin is usually less than 5 microvolts. This pattern may be accompanied by brief bursts of higher voltage spikes, which appear as single spikes above the background activity (Fig. 14.3A). This appearance is sometimes called "burst suppression". A severely abnormal trace is indicative of severe encephalopathy and is frequently accompanied by seizure activity.

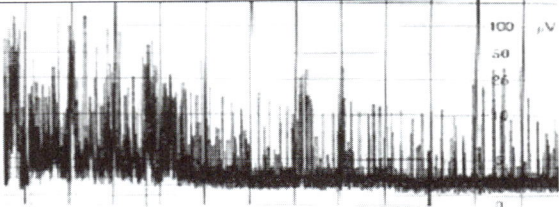

Fig. 14.3B: Repetitive seizures on aEEG, can be confirmed by raw EEG trace

Seizures

Two main seizure patterns may be seen on an aEEG recording (Figs 14.3B, 14.4 and 14.5).
1. There is a sudden rise and narrowing of the trace (reflecting the increase in EEG voltage) (Fig. 14.4).

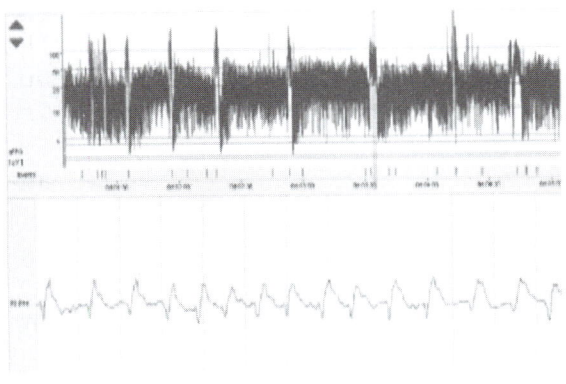

Fig. 14.3A: Severely abnormal showing a narrow low voltage tracing with lower margin below 5 microvolts and bursts of spike activity

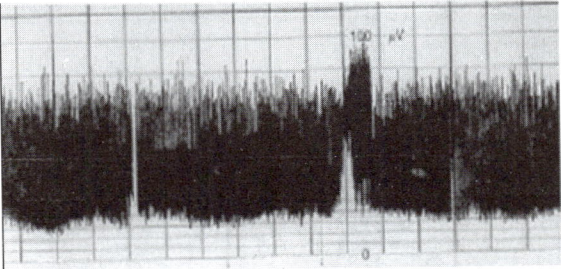

Fig. 14.4: Sudden rise and narrowing of the trace during a seizure

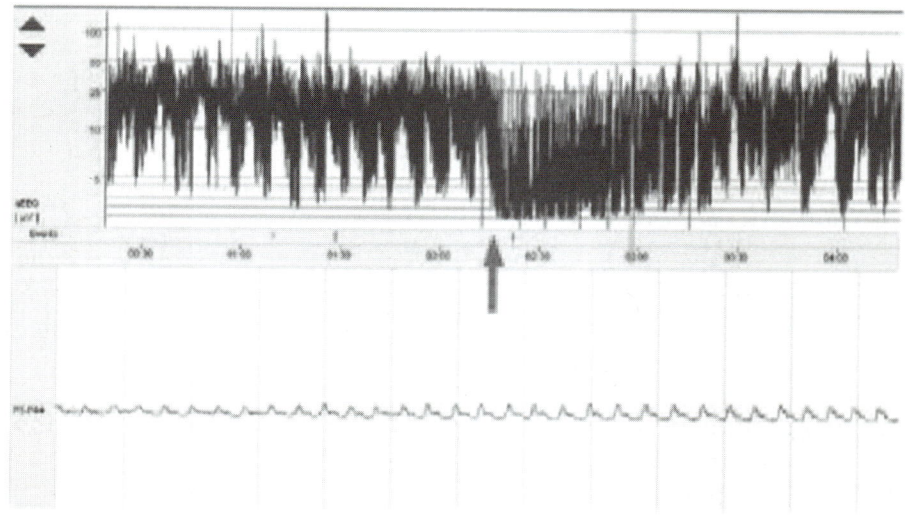

Fig. 14.5: Note the saw-tooth appearance of the trace because of regular frequent seizures

2. A narrowing of the trace without a rise in voltage may be seen. The trace returns to the previous appearance when the seizure activity stops.

Caution!

- Seizures of shorter duration (<2–3 mins) may be missed.
- An isolated sudden change in the trace suggestive of seizure may be difficult to interpret in the absence of any clinical observation.
- Artifact due to touching the infant may cause diagnostic difficulty.

Continuous seizure activity may be difficult to recognize because the trace may appear to be of normal voltage.

POTENTIAL CLINICAL APPLICATIONS OF CFM

1. In perinatal asphyxia and hypoxic ischemic encephalopathy, its uses are:
   i. Early aEEG obtained with 3–6 hours of age can be used to identify encephalopathy early.
   ii. Classify severity of encephalopathy and select patients for therapeutic hypothermia. However, aEEG should not be used as the sole criterion for therapeutic decision to cool babies (Table 14.1).

TABLE 14.1: Classification of aEEG patterns based on modified system proposed by Hellstrom-Westas

Description (μV)	Amplitude/background	Grading	Interpretation/likely prognosis
Upper margin >10 μV Lower margin >5 μV	Normal	1	1—normal long-term outcome
Upper margin >10 μV Lower margin <5 μV	Moderately abnormal	2	2, 2S—moderate HIE
Upper margin <10 μV	Severely abnormal	3	3,3S—severe HIE and poor prognosis likely
Sudden rise and narrowing of the trace or an isolated narrowing of the trace	Seizure	S	1S—isolated neonatal seizures and no encephalopathy

Background pattern	Lower margin	Upper margin	Comment
Continuous	>5 µV	>10–25 µV	
Discontinuous pattern	<5 µV	>10 µV	
Burst suppression	<5 µV		Bursts have amplitude >25 µV
Low voltage	<5 µV	<5 µV	Some variability present
Flat trace	<5 µV	<5 µV	Isoelectric trace

Normal pattern: Continuous with no seizures.
Abnormal pattern: Discontinuous, burst suppression, low voltage and flat trace.

iii. Take treatment decisions like initiate cooling therapy in asphyxia.
iv. Monitor response to anticonvulsants.
v. Prognosticate aEEG obtained between 36 and 48 hours can be used to predict neurodevelopmental outcome. A normalization of background by 48 hours is predictive of good outcome.
2. To detect and treat neonatal seizures.
3. Monitor drug effect.
4. To follow the degree of metabolic encephalopathy.
5. Determining the need for further investigations, monitoring and follow up.

Conclusions

CFM (aEEG monitor) allows the recording of cerebral activity in a continuous fashion (Fig. 14.6). It has been found useful in monitoring neonates with hypoxic ischemic encephalopathy. It can also help to identify neonatal seizures. In the coming years it is likely to be an integral part of monitoring of an encephalopathic neonate (Table 14.2).

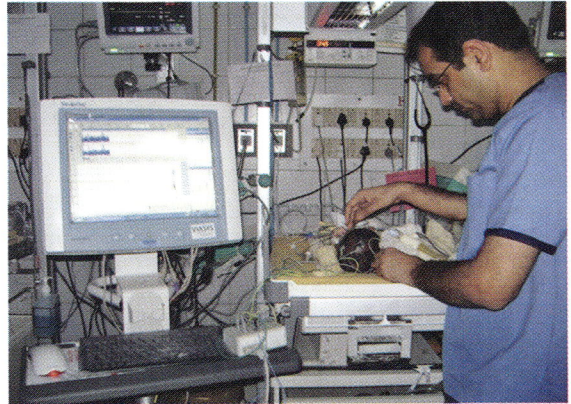

Fig. 14.6: Cerebral function monitor

TABLE 14.2: Some brands of CFMs in the Indian market

S.No.	Make	Principals	Dealers
1.	Brain Z BRM3, neonatal brain monitor	GE Healthcare Natus Healthcare	Wipro GE Healthcare Pvt. Ltd.
2.	Olympic CFM 6000 monitor	Natus Medical Incorporated	Hospimedica International Ltd.
3.	Nicolet one monitor—aEEG brain monitor	Care fusion (Viasys Healthcare)	Rohanika Electronics and Med Sys.
4.	Nemus Z	EB Neuro, Italy	Rohanika
5.	Unique CFM aEEG	Inspiration, UK	Cardiocare
6.	Neuron spectrum	Neurosoft, Russia	Chroma NV
7.	EEG 91K with aEEG	Nihon Khoden, Japan	Star Healthcare

Frequently Asked Questions (FAQs)

Q. 1. What is CFM?
The CFM (cerebral function monitor) is a device which records a single channel of EEG in a continuous fashion. By this it reflects the cerebral function in real time.

Q. 2. In what situations it is useful to monitor neonates?
It is particularly useful in neonates with perinatal asphyxia. It helps to assess the severity of insult early, take treatment decisions, monitor response and prognosticate.
- To detect and treat neonatal seizures.
- Monitor effect of drugs.

Q. 3. Does it replace the conventional EEG in neonates?
No it does not. In fact the greater recognition of abnormal patterns may increase the requests for conventional EEG.

Q. 4. Does one need to know conventional EEG to interpret aEEG recordings?
No. It does not require knowledge of conventional EEG or neurophysiology. Interpretation of aEEG can be easily learnt and taught to paramedical staff as well. It mainly involves recognition of the few normal and abnormal patterns which can learnt easily.

Q. 5. What all one should note while seeing an aEEG record?
a. Amplitudes of upper and lower margins of the trace (recorded in µV)
b. Presence of sleep wake cycle
c. Variability—*see* whether the trace is narrow or broad
d. *Background activity*
  i. Continuous (normal)—has a dark central band of activity, with normal amplitudes.
  ii. Discontinuous (abnormal)—has a wide trace with no central band of activity.
e. Presence of seizure activity.

Q. 6. Is there an option to review past events on a CFM?
The CFM machine comes with a memory storage capacity of variable duration. Old events can be traced, viewed and printed easily.

Q. 7. What are the common pitfalls in interpretation of aEEG?
While interpreting records of aEEG, keep the following pitfalls in mind.

If background activity appears elevated, it may be due to:
  i. Handling of the baby
  ii. Muscle activity (more for frontal electrodes)
  iii. High frequency ventilation
  iv. Status epilepticus
  v. Gasp artifact
  vi. ECG artifact
  vii. Cannot detect a short, focal or low amplitude seizures.

If background activity appears depressed, it may be due to
  i. Severe scalp edema
  ii. Deep sedation
  iii. Leads too close to each other.

Q. 8. How can I avoid errors in interpreting aEEG trace?
To accurately interpret aEEG
  i. Mark any care/procedures/movements or other events on the record.
  ii. Review drugs and other therapies given to the neonate.
  iii. Look at the simultaneous EEG display.
  iv. Also look for simultaneous clinical events.

Q. 9. How do I ensure a good aEEG recording?

Place the electrodes carefully. Check for impedance while placing electrodes and regularly thereafter. The impedance indicates the contact between the electrode and the patient. A loose contact or high impedance reduces the quality of the record. Keep the impedance <20 W, ideally closer to zero.

Q. 10. What is the current cost of a unit of CFM in Indian market?

Although the cost varies from brand to brand, but the approximate cost is around fifteen to twenty five lakh.

Chapter 15

BERA Phone: Automated Auditory Brainstem Evoked Response Audiometry

INTRODUCTION

The incidence of congenital hearing loss is estimated to be one out of every 1,000 live births; however, emerging data indicate that the incidence may be closer to two or three per 1,000 live births. The incidence is much higher in at-risk population (1–2%). If congenital hearing loss is not recognized and managed at an early age, the child's speech, language, and cognitive development are often delayed. Even a relatively mild hearing loss of 35–40 dB means the child misses approximately 50% of daily conversations with further consequences. The US preventive services task force has recommended screening for hearing loss in all newborn infants. Conventional auditory brainstem response (ABR) is considered to be the most reliable method for assessment of the hearing. The conventional ABR method however, is not widely used for screening because it is time-consuming and needs a well-qualified technician and audiologist to perform the test and evaluate the results. The automated auditory brainstem response (AABR) measurement has been developed as a quick and effective means of screening neonatal hearing.

Principles of Working

The purpose of a screening test is to determine if the newborn has the possibility of a significant hearing loss (>35 dB) and needs further evaluation. AABR tests the presence or absence of wave V of ABR at stimulus of 35 dB. No operator interpretation is required. The AABR uses 35 dB near hearing level click stimuli, presented mono-aurally at a rate of 93 clicks/second. The clicks have an acoustic spectrum, which is flat from 750 to 5000 Hz. After artifact rejection for ambient noise and myogenic activity, an internally programmed template-matching algorithm measures on-going electroencephalographic activity for the presence or absence of the ABR. This sampling uses a statistical test: The likelihood ratio. After reaching a likelihood ratio of 160, the machine stops collecting data and displays a "PASS" for the ear being tested. This indicates that the data collected were sufficient to discriminate between the presence of a "response plus noise" and the presence of pure noise, or a "no response" condition at better than the 99.80% confidence level. If the pass criteria are not achieved, the test is stopped with the test result "REFER". AABR hearing screening is safe, simple to operate, quick to administer to large populations and can be used by personnel who have no special audiological training. Studies have found sensitivity of 100%, specificity of 96–98%, and positive predictive value of 19%.

THE BERA PHONE

The BERA phone consists of a hand-held headphone unit (applicator), which integrates the preamplifier and a set of three fixed touch

TABLE 15.1: Common brands of BERA phone in Indian market

S.No.	Make	Manufacturer	Dealers	Unit cost (₹)
1.	MB11 with BERA phone	MAICO diagnostic	Care International	6,50,000
2.	Emissia with ABR scanner	—	Recorders and Medical Systems	4,50,000
3.	GSI audioscanner plus	Viasys	Rohanika Electronic and Medical Systems	5,00,000
4.	Neuro-audioscreen	Neurosoft, Russia	Medilife Technologies	5,00,000
5.	Viking quest incorporated	Natus Medical	Arena Medical Care	5,00,000

electrodes (Fig. 15.1 and Table 15.1). There is no need to attach disposable electrodes. The BERA phone applicator has to be held against the baby's head after the contact points on the head have been rendered more conductive by the application of electrode gel. A USB-cable connects the BERA phone with a USB port of a notebook PC or desktop computer. The result, i.e. "PASS" or "REFER" is displayed on the computer.

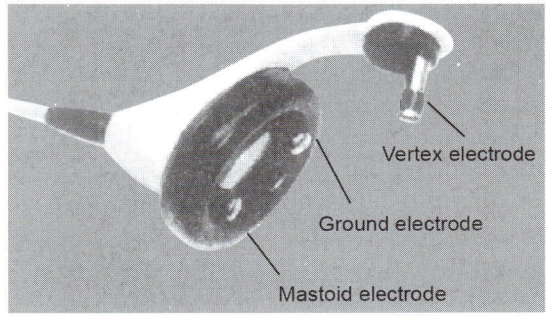

Fig. 15.1: BERA phone with integrated electrodes

Features and Specification

- Fast and automatic ABR—screening and reliable results within seconds.
- Unique BERA phone with integrated electrodes.
- CE-Chirp—stimulus ensures fast results.
- Automatic impedance check indicates impedance conditions.
- Export function of test data for quality ensuring tracking.
- USB connection for power supply and data transfer.
- Stimulation level at 35 dBHL.
- Optional "Follow-up" with standard ABR and time step stimulus.

Procedure

The BERA phone should be operated in a quiet room, so that the examinations are not influenced by outside acoustic noises. *The baby should be asleep or be very calm and not hungry.* The baby's scalp skin needs to be cleaned and prepared as proper AABR measurements require a low skin-electrode resistance (electrode impedance). If the baby's head has been lotioned/oiled or treated with any greasy skin care products in the region where the electrodes should contact then the lotion/oil must be removed carefully prior to further preparation. Preparation is done by gently rubbing the areas with electrode gel. The mastoid electrode is placed below the ear lobe, the ground electrode above the earlobe, and the vertex electrode in a straight line higher up in the direction of the forehead or vertex (Figs 15.2 and 15.3). Depending on the size of the head, the position of the vertex electrode can be adapted by rotating the black disc in which the electrode is mounted. *Do not put pressure with the BERA phone on the head.*

Neonatal Equipment

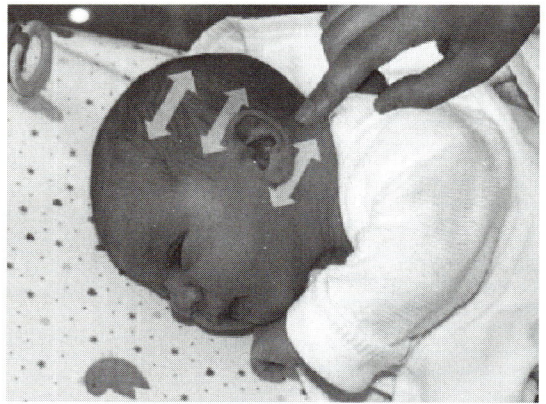

Fig. 15.2: Application of gel in electrode contact areas

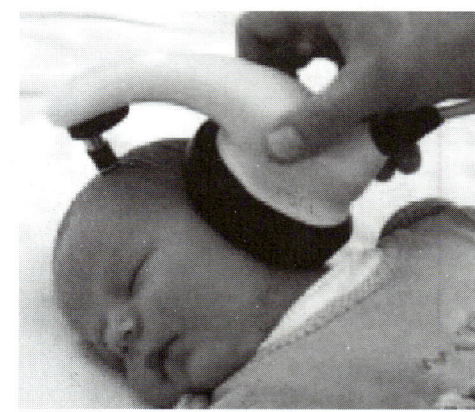

Fig. 15.3: Application of BERA phone on a baby

The correct position of the BERA phone is checked with the impedance test. The impedance is the resistance between the measuring electrodes (vertex, mastoid) and the ground electrode. This impedance is influenced by the resistance of the electrodes of the BERA phone, and more important, the resistance of the skin. The impedance should be in the range of 250 Ohm to 10,000 Ohm for each electrode pair (mastoid/ground and vertex/ground). After passing the impedance test, the measurement starts and the impedance display changes into the signal quality display. During the test, the indication line for pass criteria in the diagram continues to move upward on the graph until the green area is reached. Then 100% of the pass criteria is fulfilled and the test was passed successfully. The result "PASS" is shown in the green area (Fig. 15.4). However, a "PASS" result using this instrument is not a guarantee that the full auditory system is normal. Thus, a passing result should not be allowed to override other indications that hearing is not normal. A full audiologic evaluation should be administered if concerns about normal hearing persist.

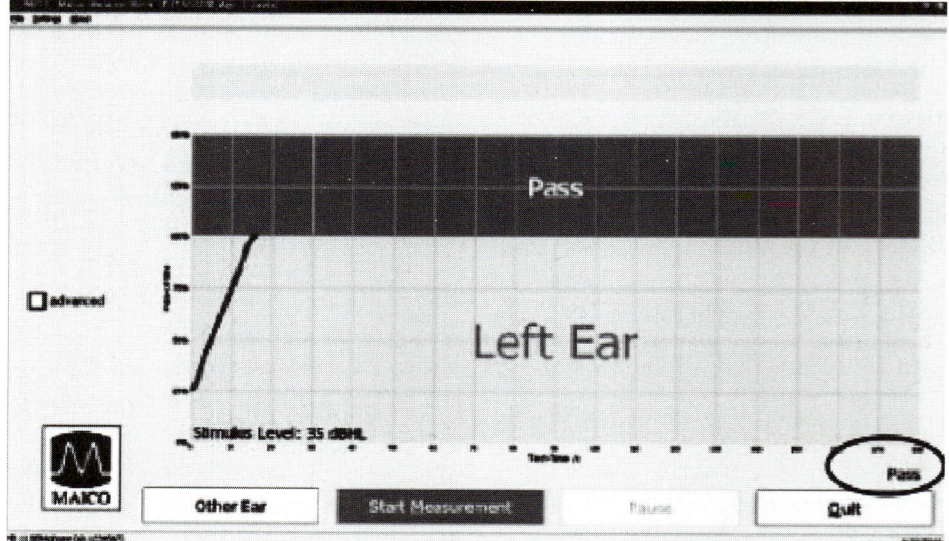

Fig. 15.4: Test result "PASS" in left ear

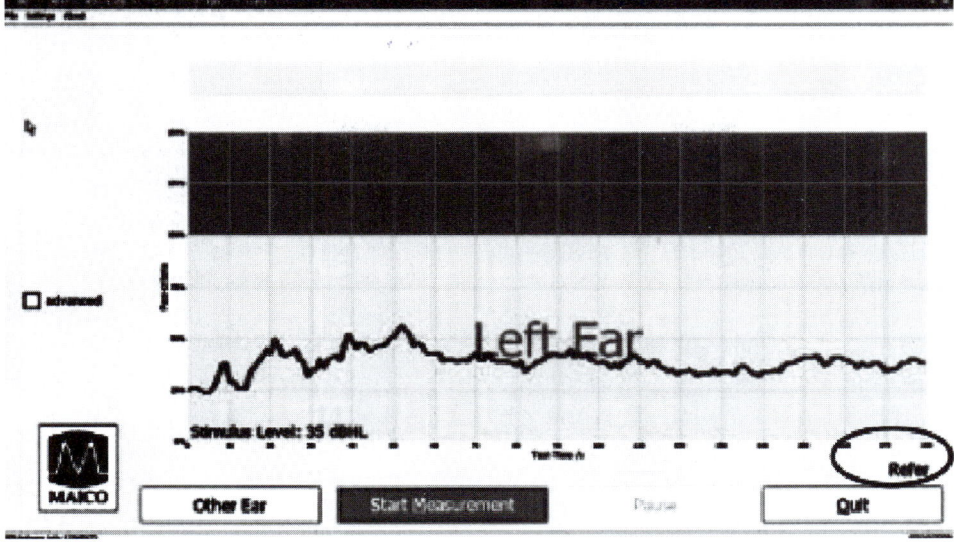

Fig. 15.5: Test result "REFER"

If 100% of the pass criteria is not reached after 180 seconds of test time, it is possible that the tested ear has a hearing loss of 35 dB or more. The small status window in the lower right of the screen shows the test result "REFER" (Fig. 15.5). If you cannot exclude bad test conditions as a reason for the "REFER", you should retest. The procedure is then repeated for the other ear.

Conclusion

For the optimal development of a newborn child, universal neonatal hearing screening should be the ultimate goal and all neonates with hearing impairment should be allowed to benefit from early diagnosis and intervention. AABR is a high-quality neonatal hearing screening method. It is easy to perform and does not require special audiometry training. It is suitable for NICU graduates before discharge as well as for healthy newborns.

Frequently Asked Questions (FAQs)

Q. 1. How does AABR compare with other methods of neonatal hearing screen such as otoacoustic emissions (OAE)?

One of the main advantages of AABR screening is that it provides information not only on conductive hearing loss and cochlear pathology, as does OAE screening, but also on the more central auditory pathology up to the midbrain. Sensitivity of OAE in the NICU population has been reported be 80%, with specificity of 92%. OAE testing alone misses auditory neuropathy/auditory dyssynchrony (AN/AD). AN/AD is a form of hearing impairment characterized by normal or near-normal hearing sensitivity but reduced auditory perceptual or speech-processing skills and electrophysiologically characterized by an absent or a grossly abnormal ABR but normal evoked OAE. The incidence of AN/AD in NICU infants has been reported to be 24%.

Q. 2. Can AABR be used in the NICU setting?
AABR is the method of choice for neonatal hearing screening in NICU settings. It is feasible for use on the ward and in the incubator, even during nasal CPAP oxygen therapy or artificial ventilation, without disturbance from either ambient noise or technical equipment.

Q. 3. Can AABR be used in preterms?

The click-evoked ABR typically appears during the 27th wk of gestation, but may occur as early as the 25th wk. Development of the infant ABR is usually complete by the second year of life. Use of AABR is officially recommended from 34 wk of gestation.

Q. 4. What precautions need to be observed while using the equipment?

The BERA phone must never be applied with pressure. The earphone with the black ear cushion must be positioned over the ear. The electrodes should not be placed in the ear canal. All electrodes must contact the skin well. Care must be taken that the mastoid electrode really remains positioned below the earlobe. Ensure quiet surrounding.

Q. 5. What are the possible sources of error in BERA phone and what are remedial actions?

1. *Bad EEG quality*: There is not enough electrode gel used.
   - All the three electrodes are not in contact.
   - There is outer noise which is interfering with the measurement.
   - The grounding is not proper which can be due to the desktop/notebook which you are using, can use battery mode.
   - The gel of the three points should not be in contact with each other as it will disturb the readings.
2. *Baby wakes up or is restless*: If baby is restless, wait with the start of the measurement until the baby calms down, or repeat the measurement when baby sleeps.

Chapter 16

Otoacoustic Emission (OAE) Machine

INTRODUCTION

Otoacoustic emissions (OAEs) are low level sounds originating within the cochlea due to outer hair cell motility. They are detectable only when both the cochlea and middle ear systems are functioning normally. They can be recorded quickly and reliably in newborns and infants with normal outer, middle, and inner ear function and hence are widely used in pediatric audiologic assessment. As with the automated brainstem response (ABR), OAEs do not measure "hearing", but rather provide a sensitive indicator of peripheral auditory function.

How are otoacoustic emissions generated?

The ear consists of a sound conducting system and a sound transducing system. The conducting system consists of the outer ear (pinna and ear canal) and the middle ear (tympanic membrane and ossicular chain). The inner ear, or cochlea, converts the mechanical impulse of vibrations transmitted to the perilymph via the ossicular chain into a nervous impulse, which is then taken to the brain where it is perceived as sound.

The inner hair cells of cochlea generate nerve impulses in response to vibrations of the basilar membrane. When sound is of low intensity, the outer hair cells help in amplification. The basilar membrane movement stimulates the outer hair cells and the cilia attached to them. This produces additional movement of the membranous labyrinth, which is conveyed to the inner hair cells. If the outer hair cells are damaged, they no longer contract in response to slight sounds and the inner hair cells are not stimulated. This produces a hearing loss for low intensity sound. If the sound is more intense, the inner hair cells are stimulated directly and they respond normally so that the ability to hear louder sounds remain unimpaired.

This active amplifier mechanism of the outer hair cells also leads to the phenomenon of sound wave vibrations being emitted from the cochlea back into the ear canal through the middle ear (otoacoustic emissions) where it can be picked up by a microphone.

What are the types of OAE?

Currently two types of evoked OAE measurements are used for newborn hearing screening: Transient evoked OAE (TEOAE) and distortion product OAE (DPOAE) (Table 16.1).

i. *TEOAE*: OAE response to a transient stimulus such as an acoustic click is measured. The response is split into frequency bands and separate responses from different parts of the cochlea are recorded. They are strongest and easiest to record in the primary speech frequency band, 1–4 KHz. Repeated click stimulation and signal averaging is used to overcome background noise and recover the emission.

ii. *Distortion product OAE*: OAE response to two pure tones of slightly different

TABLE 16.1: Transient and distortion product OAE: Stimulus and response

Type of OAE	Stimulus	What is recorded?	Normal response
Transient evoked OAE	Click has intensity of 50 dB above threshold with a peak intensity of 80–85 dB sound pressure level (SPL)	The recording is delayed 2.5 ms after the stimulus to avoid including the stimulus or the acoustic response of the middle ear in the measurement. TEOAE responses are then analyzed by frequency in order to confirm outer hair cell function along the length of the cochlea	Healthy infant ears typically produce strong OAE levels of 15 dB SPL to more than 30 dB SPL
Distortion product OAE	2 pure tones frequencies, f_1 and f_2 related with the ratio $f_2/f_1 = 1.2$	The response measured in the ear canal will contain several tones that are not present in the eliciting stimuli. These additional tones are called distortion products (DPs) and are attributable to the nonlinear distortion of the basilar membrane and is measured most commonly at the frequency given by the equation $2f_1 - f_2$	The intensity of one particular component $2f_1 - f_2$ is used as an indicator of cochlear status, plotted as a function of frequency in the 'DP-gram'. DPOAE generation is much reduced and usually absent if there is significant sensory hearing loss

frequency continuously presented to the ear is measured, the lower frequency tone f_1 applied earlier than the higher frequency tone f_2. Then a DP gram is constructed. Stimulus frequency separation $f_2:f_1 = 1.2$ results in the strongest DPOAEs.

How is OAE measured?

OAEs are recordable by means of an ear probe that is deeply inserted into the ear canal of the infant. The probe delivers a click or a pure tone and the microphone measures the OAE generated by the cochlea. In the TEOAE technique, a short click sound is delivered to the ear and the probe microphone records ear canal sounds for up to 20 ms afterward (Fig. 16.1).

The OAE probe has ports for stimulus production and a microphone transducer. The tip of the probe is disposable. DPOAE probes have an additional stimulus port for the production of 2 pure tones. The probe needs to be deeply inserted in the ear canal for maximum OAE capture and noise exclusion, with the cable positioned so as to avoid noise production on movement.

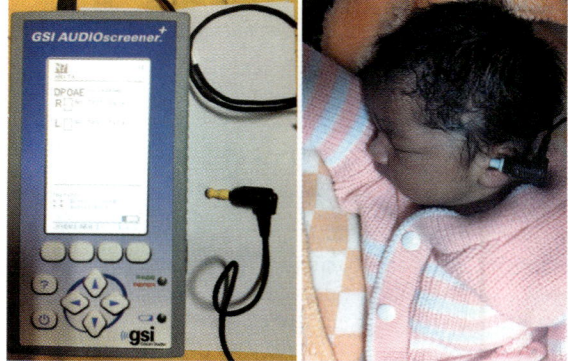

Fig. 16.1: An OAE machine with probe

OUTPUT OF OAE MACHINE AND ITS INTERPRETATION

There are different ways of recording and displaying the OAE result (Table 16.2). In devices that are used for screening, the calculations are done by the machine itself and the device simply displays a PASS or FAIL. Some devices go a step further and show OAE level, noise floor level, signal to noise ratio and a 'pass' or 'fail' verdict independently for different frequencies. Figure 16.2A shows the stimulus in the inset and the recording that begins 2.5–3 milliseconds after the stimulus. Figure 16.2B shows the analysis of recordings.

The white bars indicate signal and the dark bars indicate noise. The signal to noise ratio can be calculated at each frequency range. If it is adequate, the machine displays the result or 'PASS'.

DISADVANTAGES

Otoacoustic emissions cannot be used to screen for neural hearing loss because pathology in this disorder involves the inner hair cells, eighth cranial nerve, and brainstem with intact outer hair cells. So infants with neural hearing loss will fail ABR but pass OAE. The Joint Committee on Infant Hearing 2007 states that infants cared for in the NICU for greater than 5 days are at highest risk for neural hearing loss and, therefore, should be screened only with AABR. Some hospitals use a two-step screen with both AABR and OAE.

TABLE 16.2: Common models of OAE available in Indian market

Model name	Local dealers	Principals
Nicolet's Viking	Rohanika	Viasys
Audxpro; portable DPOAE and/or TEOAE	Natus	Biologic AuDX (Natus)
Gsicorti™	—	Grason-Stadler, US
Eroscan	—	Maico Diagnostics, US
Otocheck	—	Otodynamics Ltd., US
Otoread	—	Interacoustics, US

Note: Various models available in the market (approx. 3,00,000/–)

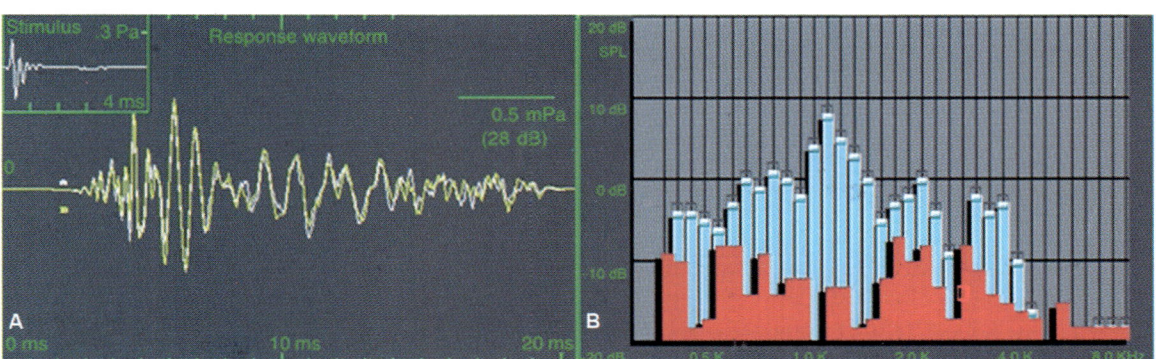

Fig. 16.2: Transient-evoked otoacoustic emission (TEOAE) in a normal ear with intact middle and cochlear function. (A) The stimulus is shown in the inset. The recording starts after a brief time of 2.5–3 milliseconds following the stimulus. (B) Analysis of the recording is shown on the right. The upper blue columns indicate the signals at each frequency for this ear. The red bars indicate noise. If there is a sensory hearing loss at any frequency, the signal will not be obtained

Frequently Asked Questions (FAQs)

Q. 1. How is OAE different from AABR (automated auditory brainstem response) audiometry?

OAE	AABR
Measures OAE generated by sound	Measures brain response to sound
Identify cochlear pathology	Identify cochlear and retro-cochlear pathology
Can identify till moderate hearing loss	Can identify severe and profound hearing loss
Highly affected by external and middle ear status and environmental noise	Less affected by external and middle ear status
Little signal processing is required to extract these responses from noise and fully validate frequency specific can be made in a few seconds	Requires electrodes and must be extracted from the relatively strong EEG background signal over a longer period of signal averaging

Q. 2. What are the causes of false positive OAE?

A false positive result occurs when the test fails in spite of a normal hearing. It may result from outer or middle ear dysfunction, including the presence of a transient conductive hearing loss (fluid or debris) or noise interference. It can also occur if there is poor probe placement.

Q. 3. Is OAE machine essential in every NICU?

Screening all newborns for hearing loss and providing them with rehabilitation as soon as indicated are advocated by professional organizations worldwide (Joint Committee on Infant Hearing [JCIH], 2007; National Institutes of Health [NIH], 1993; World Health Organization [WHO], 2010). Hearing screening is imperative in all NICUs. Whether the unit is using AABR or OAE or a two-tier system (TOAE followed by AABR) has its own advantages and disadvantages.

Q. 4. Does a normal OAE rule out hearing loss?

OAE is a simple, non-invasive, short duration and cost effective method and is a suitable test for neonatal hearing screening. Even though nearly two-thirds of patients were detected by this method, a negative predictive value of nearly 95% makes it a good screening test to rule out hearing loss.

Q. 5. In high-risk newborn is OAE alone a sufficient hearing screen test?

Perhaps no; the true sensitivity and specificity of newborn hearing screening are difficult to estimate from most screening programs. One-stage screening with an ABR or OAE test can detect 80 to 95% of affected ears, depending on how an abnormal test result is defined. The two-stage protocol of OAE and ABR still misses 11% of affected ears, but is more specific than testing with the ABR or OAE alone. Additionally, OAE is highly affected by external and middle ear status and environmental noise and hence cannot be a sufficient hearing screening test. However, more research is required before a decision of the tool to be used for universal newborn hearing screening is made.

Q. 6. How should the machine be maintained?

OAE has a screening unit, a probe assembly and a disposable probe. Some precautions are required for the maintenance of this equipment.

 i. The screening unit is a sensitive piece of electronic equipment. It should not be dropped or left in an extreme hot or cold temperature or in a humid environment.
 ii. The probe assembly is an expensive part that can be damaged easily so it has to be handled carefully. The connecting pins should not bend. The probe should be left attached to the screening unit rather than detaching it after each use. Probes also

have an inexpensive replaceable nozzle that may need to be changed occasionally, particularly if it gets clogged with wax. Once a nozzle has been removed, one would typically need to replace it with a new one that fits tightly at the base.

iii. The disposable probe tips or covers are designed to fit snugly on the probe and form a seal in the child's ear canal. One should use a new probe cover for each child being screened and discard them after they are used.

Q. 7. What are the pre-requisites for obtaining OAE?

i. A quiet infant and quiet environment.
ii. Unobstructed external ear.
iii. Adequate seal of the ear canal with probe.
iv. Absence of middle ear pathology like effusion.

Q. 8. Which out of AABR or OAE is the preferred technique for newborn screening?

Newborn hearing screening programs in which clinical decisions are made after a single screening test have the highest overall referral rates. Programs that repeat the same screening test on those newborns who do not pass an initial test have lower referral rates than the one stage screening design. Still other programs use OAEs for the initial screening, followed by an ABR for those newborns who do not pass the OAE test. These combination-screening programs tend to have low referral rates and low overall costs.

A recent study comparing two-step TEOAEs and BERA found that BERA was more effective for newborn hearing screening because it yields fewer false positive results and a lower referral rate compared with TEOAE, resulting in a smaller percentage of infants lost during follow-up.

Section D. Equipment for Therapy/Treatment Purposes

17. Phototherapy Units
18. Infusion Pumps
19. Cooling Equipment for Therapeutic Hypothermia

Chapter 17

Phototherapy Units

Phototherapy has been in use since 1958 and has stood the test of time. Despite the use of phototherapy for over 40 years now, there are still many simple issues, which the treating physician should know. Phototherapy involves exposure of the skin of the jaundiced baby to blue or cool white light of wavelength 400–520 nm. Detoxification begins immediately by the production of configurational and structural photoisomers of bilirubin in the skin and precedes the fall in serum bilirubin. Special lamps emitting light predominantly in these wavelengths are considered to be the most effective and specific for administering phototherapy. Light is effective in the treatment of hyperbilirubinemia mainly because of its blue content. Sunlight is relatively ineffective despite its ability to bleach the infant's skin because its blue content is low. Besides, hyperpyrexia and skin burns may occur due to prolonged sunlight exposure.

MECHANISM OF PHOTOTHERAPY

Phototherapy reduces the serum concentration of bilirubin and the risk of bilirubin toxicity. It has been found to be effective in treating hyperbilirubinemia in hemolytic as well as in non-hemolytic settings. This has dramatically reduced the need for exchange transfusion. Unconjugated bilirubin in skin gets converted into water-soluble photoproducts on exposure to light of a particular wavelength (400–520 mm). These photoproducts are *water soluble, nontoxic* and *excreted through the intestine and in the urine.* Phototherapy acts upon bilirubin bound in the skin and subcutaneous tissue up to a depth of around 2 mm. For phototherapy to be effective bilirubin needs to be present in skin, hence there is no role for prophylactic phototherapy.

a. *Configurational isomerization*: The normal Z-isomers of bilirubin are converted into yellow E-isomers. (Z and E are chemical terms, akin to the terms *cis* and *trans*, that denote the stereochemistry of double bonds.) Although this reaction is instantaneous upon exposure to light, the clearance of *E*-isomers is slow. These photoisomers revert to native bilirubin in bile. After exposure to 8–12 hrs of phototherapy, depending on the dosage and spectral quality of the light used, Z-isomers may constitute 25% of the total serum bilirubin (TSB) which in turn will be converted into less toxic E-isomers within a few hours.

b. *Structural isomerization*: This is a relatively slow, but irreversible reaction whereby bilirubin is converted into another yellow isomer, lumirubin, which is excreted rapidly. The formation of lumirubin is directly proportional to the dose of phototherapy.

c. *Photo-oxidation*: This is an even slower reaction which leads to colorless water-soluble photoproducts that are excreted in urine.

The relative contributions of these different mechanisms to the overall elimination of bilirubin are not known.

The effectiveness of phototherapy depends upon

1. The level of initial bilirubin.
2. Area of skin exposed. Body surface area exposed to phototherapy determines the efficacy. Therefore, babies exposed to double surface phototherapy are reported to have faster decline of bilirubin levels.
3. The dose of light (measured as irradiance X duration of treatment X% BSA treated).

There is a dose-response relationship; with increasing dosage of light, the fall of bilirubin is faster. The rate depends on the spectrum of light delivered and the irradiance, which is affected by the distance of light source from the baby. The farther the phototherapy unit is from the patient, the less will be the irradiance delivered. Halogen lights should not be placed close to the baby as they deliver considerable heat. Light emitting diode (LED) lights emit a little heat, and like fluorescent lights, can be brought close (up to 10 cm) to the baby.

Increasing the surface area exposed can be achieved by using a light source above and below the infant or by placing reflecting material around the inside of the bassinette or incubator.

INDICATIONS FOR PHOTOTHERAPY

Simply put, phototherapy should be initiated whenever it appears that bilirubin could reach such levels that can cause bilirubin induced neuronal damage (BIND). The question is how does one predict that? In actual life, there are a number of variables which determine the rate of rise of bilirubin and the susceptibility to BIND. The vulnerability of the brain for BIND increases with immaturity, acidosis, asphyxia, higher free bilirubin levels, etc.

Each institution must create its own guidelines. For an individual case, here are a few general guidelines that the pediatrician must be aware of before deciding to start phototherapy:

1. For term healthy babies, American Academy of Pediatrics guidelines can be followed.
2. Guidelines are provided for very low birth weight babies. As a rough guide, phototherapy is indicated at a level equal to 1% of the body weight (for example, 10 mg/dl in a 1000 gm baby or 15 mg/dl in a 1500 gm baby). Exchange blood transfusion is warranted when the TSB level is 5 mg/dl higher than the phototherapy level. However, the overall clinical situation needs to be considered to arrive at a proper decision.
3. In case of hemolysis, start phototherapy at a lower level.
4. Acidosis, asphyxia, hypoglycemia or sepsis make the blood–brain barrier more porous to bilirubin. So, consider early phototherapy.
5. In case of prolonged jaundice (>3 wks), one should always check fractional bilirubin estimation. Phototherapy is contraindicated in the presence of conjugated hyperbilirubinemia (2 mg/dl) because it may result in bronze baby syndrome.

PHOTOTHERAPY LIGHT SOURCES

Emitted light should be filtered to remove harmful infrared and ultraviolet radiation. Light should be focused on the baby. Mobile units are preferred because they can be used for babies nursed in cots, incubators or radiant warmers. The height should be adjustable, while a few units may be tilted on axis. Phototherapy lights may be mounted on the radiant warmers themselves. A standard locally fabricated phototherapy unit should consist of 6 to 8, white fluorescent tubes (20 or 40 W), preferably of the daylight variety, if blue or special blue tubes are not available locally (Figs 17.1 and 17.2).

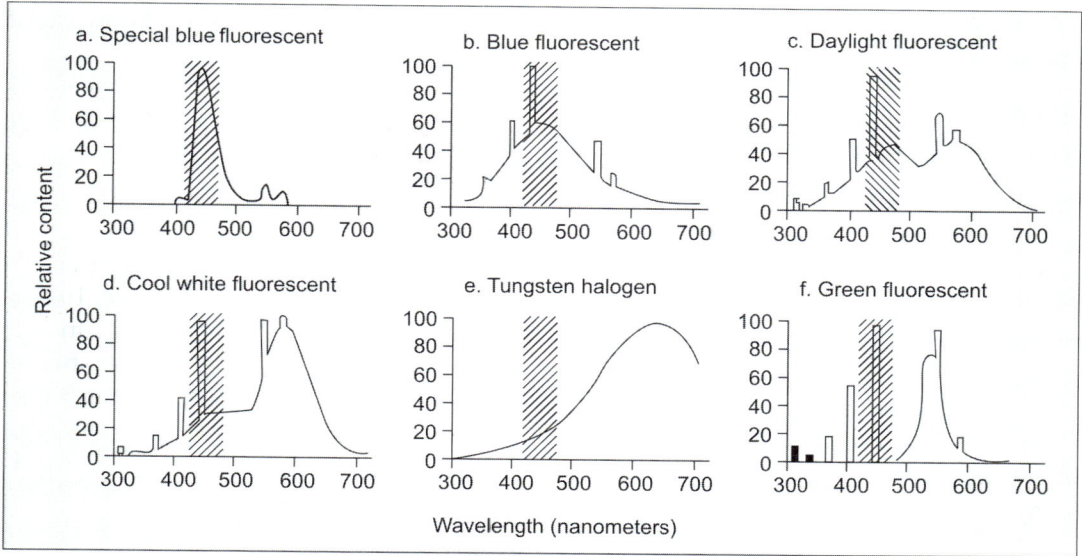

Fig. 17.1: Spectral emission ranges with different light sources [based on data from olympic medical (a to d); *Pediatrics* 1980;65:795–798 (e); *Arch Dis Child* 1988;63:461–462 (f)]

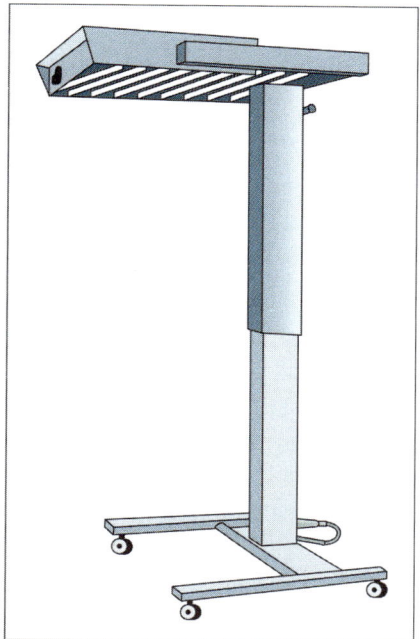

Fig. 17.2: Phototherapy unit

a. ***Halogen spotlights***: Spotlight phototherapy units generally use a 150 W, 21 V halogen bulb with a specially coated reflector which absorbs infrared wavelength. A fan continuously cools the hot bulb. Options for varying aperture diameter and different filters are available. Positioning of the light on the baby is critically important in maximizing the spotlight's effectiveness. They are most effective when located directly above the infant at a distance of 45–50 cm. Directing the light from the side of the crib or allowing the circle of light to move off the infant's largest surface area, e.g. abdomen, significantly reduces the dose delivered. A few halogen spotlights incorporate a dosimeter which depicts how much dose of phototherapy the baby has received. It considers the total irradiance received by the baby and multiplies this by the duration in hours.

b. ***Fluorescent lamp devices***: These have optimized blue light emission at 400–520 nm wavelengths. Special blue fluorescent tubes are labeled F20T12/BB or TL 20 W/52. Regular blue fluorescent tubes (F20T12/B) deliver much less irradiance. If possible, the irradiance should be measured at regular time intervals to ensure that an adequate dose is being delivered. Fluorescent tubes lose about 35–40% of blue light irradiance after 1200 hours of use. Directing the light from the side of the infant significantly reduces the dose delivered. These lights can provide an irradiance of >25–30 mW/cm^2/nm in the 400–520 nm range when placed closely, thus making phototherapy maximally effective particularly when the greatest body surface area is exposed.

c. *Fiberoptic pads*: These devices use plastic fiberoptic light guides to deliver light from a halogen lamp to illuminate a blanket or pad which is wrapped around or placed under the baby. These devices deliver light in the 400 to 550 nm spectral band. The pad is cool and can be placed in direct contact with the baby. They can be used as an auxiliary light source to increase the surface area exposed or as the sole source of phototherapy, particularly in preterm infants. In recent models, the halogen light source has been replaced by high intensity high power LED bulbs. This increases the irradiance delivered by the pads.

d. *Compact fluorescent tubes*: These are short (approx. 5 to 7 inch) double-folded tubes (9–18 W) that emit blue or white light. Several (6–8) are housed in a panel with reflectors. As they do not produce much heat the distance to baby can be relatively short thus increasing the irradiance delivered. Most of them produce an irradiance of 20–30 mW/cm^2/nm when placed close to the baby and can be used in combination with overhead lights to increase the surface area of exposure (Fig. 17.3).

e. *Light emitting diodes (LED)*: Blue LED devices emit a narrow spectrum 460–470 nm that overlaps the absorption spectrum of bilirubin. They are power-efficient, portable devices with low heat production that can be kept close to the baby. They are durable and long lasting with low power consumption (Figs 17.4 and 17.5).

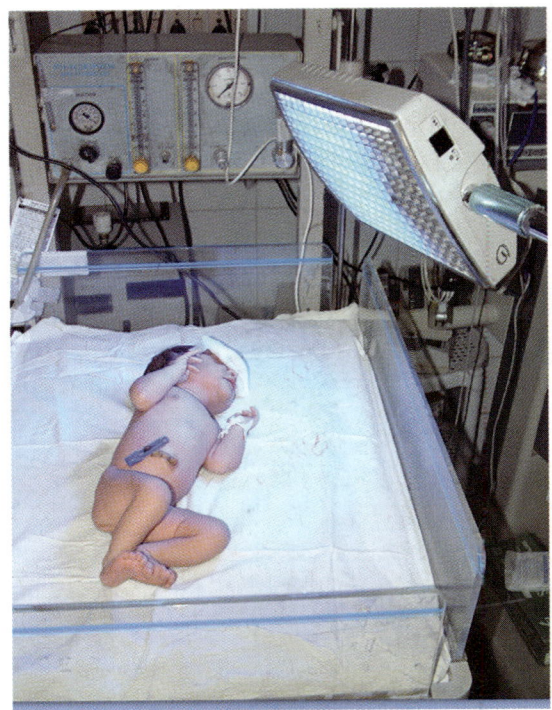

Fig. 17.3: H- or U-shaped 4–6 inch tubes 9 or 18 W compact phototherapy unit

Tips towards delivering safe and effective phototherapy

1. Protect the eyes with eye patches.
2. Keep the baby naked with a small nappy to cover the genitalia.
3. After switching on the unit check that all tubes/bulbs are on.
4. Place the baby as close to the lights as the manufacturers' instructions allow. Use white cloth or aluminum foil to reflect light back onto the baby, making sure not to impede the air flow that cools the bulbs.
5. Do not place anything on the phototherapy unit (this may block air vents or light and items may fall on the baby).
6. Encourage frequent breastfeeding. Unless there is evidence of dehydration, supplementing breastfeeding or providing IV fluids is unnecessary.
7. Change position supine to prone after each feed to expose the maximum surface area of baby to phototherapy.

8. Keep diaper area dry and clean.
9. Phototherapy does not have to be continuous and can be interrupted for feeding, clinical procedures, and to allow maternal bonding.
10. Monitor temperature every 4 hours and weight every 24 hours.

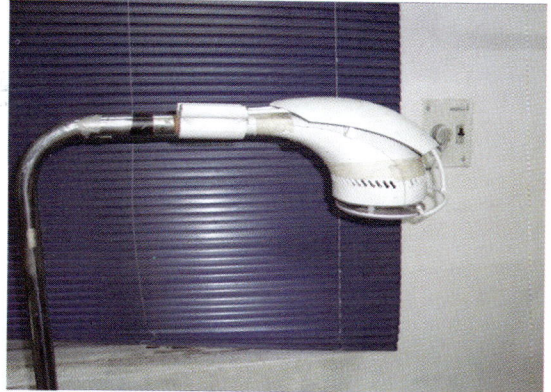

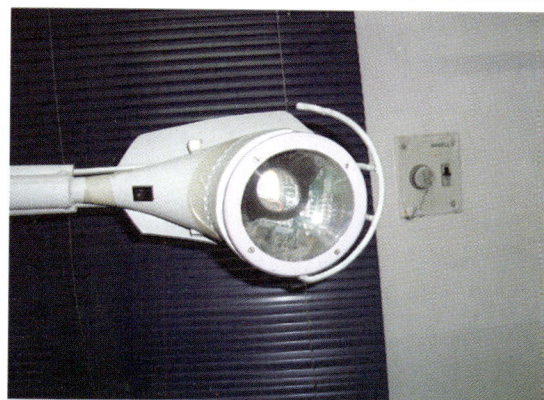

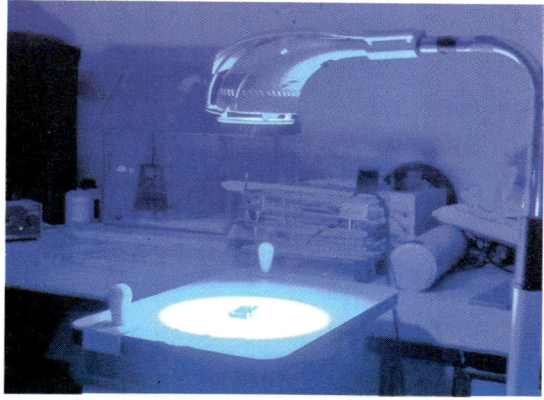

Fig. 17.4: LED lights for phototherapy

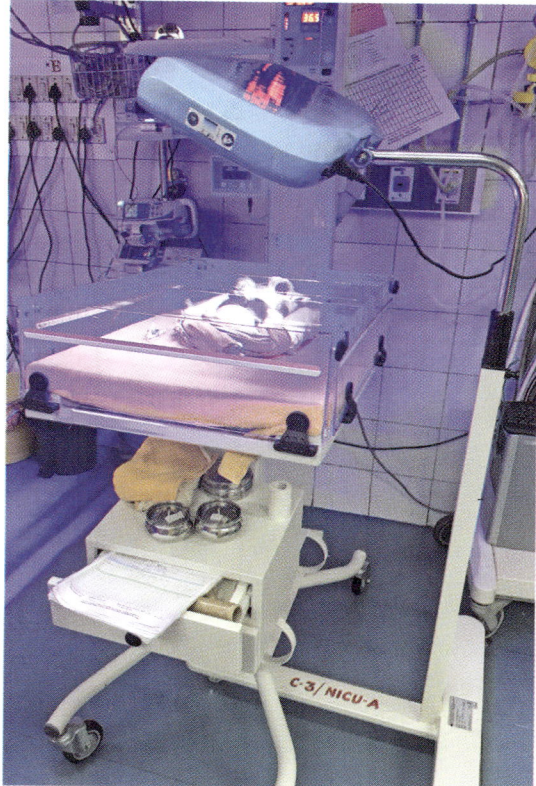

Fig. 17.5: Baby under LED phototherapy

11. Measure plasma/serum bilirubin frequently ~ every 12 hours. Visual assessment of jaundice during phototherapy is unreliable.
12. Change tube lights every 6 months (or usage time >1200 hrs) whichever is earlier; or if tube ends blacken or if tubes flicker.

Side Effects

Phototherapy is in use since the last 40 years and has an excellent safety record. The predominant adverse effects of phototherapy include rash, overheating, dehydration and diarrhea. Retinal damage is prevented by shielding the eyes. *In vitro* studies have suggested that DNA damage may be caused by phototherapy. Recently, the effect of phototherapy on cerebral blood flow velocity

(CBFV) has been reported. Phototherapy increased mean CBFV in all preterm infants, which returned to pre-therapy values after discontinuation of phototherapy only in non-ventilated babies. Even in term babies, phototherapy increased CBFV, which returned to pre-therapy level upon discontinuation of phototherapy. In addition, phototherapy influences cytokine production by peripheral mononuclear blood cells. Phototherapy also has photo-oxidative effects on intravenous lipids, proteins and drugs like amphotericin B. Phototherapy has been shown to affect short-term behavior of the term infant, which has been attributed to maternal separation. This least discussed and often overlooked aspect, is the most common side effect. So one should encourage the mother to breastfeed and interact with her baby regularly during phototherapy (Table 17.1).

TABLE 17.1: Phototherapy units available in the market

S.No.	Type of unit	Principals	Dealers	Unit cost (₹)
A.*	White fluorescent tube (6 to 8; 20 W)	*	*	20,000
	Double surface fluorescent			35,000
B.	Compact fluorescent lights (6 to 8; 21 W)	Medela	Rohit surgicals	60.000
		Phoenix	Phoenix	25,000
		Nice Neotech	SBP Mediare	20,000
		SS Technomed	Global Medical Sys	20,000
		Meditrin	Meditrin	25,000
C.*	Blue light (2 to 4; 20 W) and white (2 to 4; 20 W)	Atom	Vishal Surgicals	40,000–2,00,000
		Medela	Rohit Surgicals	
		Draeger	Draeger	
		Wyer	Rustagi Surgicals	
		Ameda	Medisphere	
		Heraeus	Medex	
		Choongwae	Global Medical	
		Phoenix	Phoenix	
D.	Halogen bulb (Single; 150 W; 21 V) Hillrom	Datex-Ohmeda	Phoenix	60,000–90,000
		Olympic	Rustagi Surgicals	
		Phoenix		
E.	Bili-blanket fiberoptic Bili Bed™	Ginevri	Global Medical System	1,00,000–2,50,000
		Wallaby	Global Medical System	
		Medela		
		Olympic	Rohit Surgicals	
		Datex-Ohmeda	Rohit Surgicals	
		Ibis Medical	Ibis Medical	
		Shrichitra	Shrichitra	40,000
			GE	60,000
		Fanem	Fanem	80,000
		Phoenix	Phoenix	40,000
				40,000
		Ibis Medical	SS Technomed Pvt Ltd.	50,000

*LED: Lullaby LED GE, LED high intensity spot (Bilitron 3006/Bilitron bed 5006/Bilitron sky 5006), LED high intensity spot, Sunshine LED, Ibis Medical

Caution

1. Do not use phototherapy without trying to find the cause of jaundice.
2. Phototherapy results in dehydration and iatrogenic hyperthermia/hypothermia.
3. Blue light may interfere with monitoring of cyanosis. Blue light causes nausea, giddiness and headache, which may disturb the staff.
4. In direct hyperbilirubinemia, phototherapy results in bronze baby syndrome (green color).

Maintenance Tips for Phototherapy Units

If possible, use a spectroradiometer (calibrated for 430–490 nm waveband) to measure the irradiance and replace the tubes or halogen light sources with new bulbs if the irradiance has fallen below 10–12 mW/cm^2/nm or to two-thirds of its original value. It is not necessary to measure irradiance before each use of phototherapy; but it is important to perform periodic checks of phototherapy units to make sure that an adequate irradiance is being delivered. Note that irradiance meter is designed to measure specific light source for precise bandwidth.

If a spectroradiometer is not available, halogen bulbs and fluorescent tubes should be changed every 1200 hours or 6 months of intermittent use, whichever is earlier, or when fluorescent bulbs flicker or tube ends blacken. Maintaining a logbook for duration of use or built-in timer in the device is useful. Keep bulbs and reflectors dust free and do not cover the unit with cloth, charts or files as this will prevent circulation of air and heat the bulbs.

Most of us do not have a flux-meter to measure blue light output of phototherapy unit. We must remember that the phototherapy unit is not just a source of a visual illumination. It is biomedical equipment which needs to be serviced by a trained and knowledgeable person. Ensure that supplier/dealer of phototherapy unit has flux-meter with him and checks the efficiency of the phototherapy unit periodically.

Conclusion

Phototherapy continues to be the preferred method of treatment for neonatal hyperbilirubinemia by virtue of its safety and non-invasive nature. Appropriate use of phototherapy has significantly reduced the need for exchange blood transfusion. It is convenient and inexpensive. Hence, it has been immensely popular even in small hospitals and nursing homes.

Frequently Asked Questions (FAQs)

Q. 1. Can one use white tubelights for providing phototherapy?

Cool daylight lamps (fluorescent tubelights 6 to 8, 20 W each) with a principal peak at 550 to 600 nm and a range of 380 to 700 nm are most commonly used phototherapy lights in India. These units provide 10–14 mW/cm^2/nm in the wavelength range of 425–475 nm, when the fluorescent tubes are new. With usage the irradiance is bound to be much lower than required for therapeutic purposes. These units are effective in the treatment of non-hemolytic jaundice in term and preterm neonates. Occasionally, these are not effective especially in cases of severe or rapidly increasing neonatal jaundice. Remember the tubes should be changed every 6 months or earlier if irradiance is being monitored. Putting a blue plastic persplex or glass sheet in front of light source will not increase irradiance of unit in blue-green range, but rather decrease it. Phototherapy with tubelights, particularly if used in adequate dosage (or, when necessary, in a double configuration) is perfectly adequate for the therapy of hyperbilirubinemia. These lights do not disturb the nursing staff and this is the only form of phototherapy in which safety has been established.

Q. 2. Does blue light provide any additional benefit for lowering bilirubin levels?

Because bilirubin is a yellow pigment, the bilirubin molecule is only capable of absorbing the photons of violet, blue, and some green. When broad-spectrum white light is used, only a fraction of the light affects the bilirubin. Blue light at approximately 450 nm is better absorbed than green light but green light (because of its longer wavelength) penetrates the skin better. The most effective lights for phototherapy are those with high energy output near the maximum adsorption peak of bilirubin (450 to 460 nm). Special blue lamps with a peak output at 425 to 475 nm are the most efficient for phototherapy and these do not emit harmful ultraviolet (UV) rays. Blue-green light may interfere with the monitoring of cyanosis. In addition, blue light causes nausea, giddiness and headache to the staff working in nursery. Green light causes erythema and subsequent tanning of skin. A combination of alternating four blue and two white tube lights (20 W each) are sufficient to provide adequate irradiance of 20–30 mW/cm^2/nm in the wavelength range of 425–475 nm.

Q. 3. What specifications of blue tube should be used for phototherapy?

Use Philips 20 W, 2 ft blue fluorescent tubes TL-52 for phototherapy units. These tubelights cost ₹ 700–900/- each. Do not use TL-02, TL-03, TL-05 philips blue fluorescent tubes. These are cheaper (cost ₹ 250/- each) but produce harmful UV rays used for sterilization, attracting insects for killing or dermotherapy for vitiligo. Imported 20 W tubes for treatment of jaundice come with following specifications: F20T12/B—regular blue; F20T12/BB—special high output; 20SBW-NU—blue white light; 20SBG-NU—blue green light.

Q. 4. What are the advantages and limitations of double surface phototherapy?

By using double surface phototherapy more irradiance can be provided to a jaundiced baby which will result in faster decline of serum bilirubin. Unfortunately, the surface on which the baby has to lie is not comfortable for the baby in locally fabricated units available in India. A convenient way of providing double surface phototherapy is using conventional blue light and undersurface fiberoptic bili-blanket phototherapy.

Q. 5. What is a halogen bulb phototherapy?

Halogen white light (150 W, 21 V) having significant output in blue spectrum is useful for treating neonatal hyperbilirubinemia. Aperture size (3–20 inch) and unit to mattress distance can be controlled. Inner reflecting surface lining of halogen bulb absorbs majority of infrared (IR) rays (a fan continuously cools it) and a UV filter in front of bulb blocks harmful UV rays reaching the baby. Indigenous halogen bulb phototherapy units may lack UV filter and thus may not be safe for use. A few warmers which have side mounted halogen bulbs (50 W, three on each side) without UV and IR filters work like an uncontrolled thermal therapy unit rather than safe phototherapy.

Q. 6. How does bili-timer work and help in monitoring of phototherapy?

Bili-timer has an "electric eye" that turns the timer on only when the phototherapy light is on. Accumulated exposure time is continuously displayed during treatment, accurate to a tenth of an hour. It is battery operated and simple to use with any unit. Accurately recording exposure time is essential to evaluate treatment and maintaining medico-legal records.

Q. 7. What are relative contraindications of using phototherapy?

1. As an isolated therapy when exchange transfusion is indicated for removal of antibodies in the presence of rapidly rising bilirubin.
2. In presence of direct hyperbilirubinemia.
3. Porphyria.
4. Concurrent therapy with tin protoporphyrin.

Q. 8. Why does bronze baby syndrome occur?

Bronze baby syndrome occurs when phototherapy is used in the presence of hepatic dysfunction and cholestasis leading to high serum porphyrins and copper. Bilirubin

iii. The instrument should not be exposed to excessive humidity, extreme heat or cold for prolonged periods.
iv. A daily check of the strip guide, reflectance disc and optical window should be made. The strip guide can be cleaned with a brush and water or a mild detergent, after removing it from the instrument. The reflectance disc and optical window can be cleaned with a soft, lint free cloth or lens tissue soaked with water, surgical spirit or alcohol.
v. The instrument should be handled gently.

How to store the strips properly?

The reagent strips contain enzymes glucose oxidase and peroxidase. Activity of these enzymes is affected by heat, humidity and excessive exposure to light. Most manufacturers recommend storage in a cool dark place at a temperature less than 25°C; but these should never be frozen. The bottles contain 'silica gel' to absorb the moisture. The color of the strip should be checked before using it.

In order to economize, many users cut the reagent strips into 2 or 3 strips, for visual reading. However, as the strips have more than one layer, this may alter the precision. As far as possible, cutting of strips should not be resorted to.

How to select a product for your unit?

This is a difficult question, as none of the manufacturers recommend use of their meters/reagent strips in neonates, because of the problems mentioned above. However, because of the requirement for rapid diagnostic method, the same have to be used, understanding the limitations well. The procedure of estimation should be simple. Visual techniques and most of the reflectance meters (glucometer, etc.) require wiping/washing of the strip after a particular period. Any error here can lead to errors in results. 'OneTouch™, meter does not require any wiping of the strip.

The meters should be preferably calibrated for plasma glucose. This may improve the precision. The reagent strips should be freely available and the cost should be reasonable. The strips should be stable for sufficient period of time in tropical climate.

Glucose estimation meters cost ₹ 3,000/- to 8,000/- each, while each strip cost varies from ₹ 8/- to ₹ 20/- .

Common Brands of Glucometer Available in India

A. Glucose Oxidase Based Reflectance Meters

1. Ames glucometer (Bayer diagnostics)
2. OneTouch™ (Johnson and Johnson)
3. Lifescan (Johnson and Johnson)
4. Glucosite (GDS diagnostics)
5. Refcolux (Boehringer Mannheim)

B. Glucose Oxidase and Electrode Based Analyzers

1. Pulsatum (Pulsatum Healthcare Pvt. Ltd.)
2. Glucometer Elite (Bayer)

C. Reagent Strips for Visual Reading

1. Dextrostix (Bayer)
2. Glucostix (Bayer)
3. Hemoglukotest (Boehringer Mannheim)

Frequently Asked Questions (FAQs)

Q. 1. Why is there a need for rapid diagnostic tests for blood glucose estimation in a neonatal ICU?

Hypoglycemia in neonates is not uncommon and can be responsible for neurological abnormalities if not detected and treated in time. A rapid bedside diagnostic method is therefore required to screen neonates at risk.

Q. 2. Are there any reagent strips available for use in neonates?

The rapid diagnostic blood glucose reagent strips were basically designed for use in diabetics. So, they cover a wide range of glucose values; however, their ability to pick up low values is poor. Infact, most of the manufacturers do not recommend the use of these strips in neonates.

Q. 3. What is the principle of reagent paper strips?

In a reagent paper strip, the blood glucose is acted upon by the enzyme 'glucose oxidase' to yield H_2O_2 which is then measured by use of a peroxidase step coupled to a colored oxygen acceptor.

Q. 4. What are the problems with use of reagent strips for detection of neonatal hypoglycemia?

High hematocrits and high viscosity of neonatal blood interfere with the estimations. They lead to discoloration of the pads and also impede the diffusion of plasma into the test pad of the strip. Bilirubin and hemoglobin also interfere. All these can falsely lower the values. Also the values in lower range are imprecise.

Q. 5. Can the estimation by reagent strips be improved upon?

The precision of the reagent strips can be improved to some extent by coupling this with a suitable reflectance meter. However, the above mentioned problems still remain. The meter reading will be more precise than visual estimates.

Q. 6. What is the sensitivity and specificity of reagent strips in detecting neonatal hypoglycemia?

Various studies show that reagent strip screening detects only about 85% of true cases of hypoglycemia and only 75% of babies who are normoglycemic. This suggests that reagent strip tests are unsuitable for diagnosing neonatal hypoglycemia. However, in absence of other cheap, easily available technique, these strips have to be relied upon (either alone or coupled with reflectance meter) in neonatal setups.

Q. 7. What are the alternatives to reagent strip tests?

Glucose electrode based automated system can be installed in the ICU. These are precise, however expensive ($15,000).

Q. 8. What precautions must be taken when using reagent strip test?

One should always keep in mind the inaccuracies of this test, so a sample should always be taken for laboratory confirmation. While taking sample for reagent strip test, the following precautions should be taken:
a. Sample should be free flowing; do not squeeze the part.
b. Avoid capillary sampling if the peripheral perfusion is poor.
c. Avoid contamination of test pad with alcohol.
d. The test pad should be completely covered with blood.
e. Carefully time various steps such as wiping or washing and the reading.
f. The strips should never be cut into 2 or 3 strips to economize.

Q. 9. How should one take care of the reflectance meters?

a. These meters should be calibrated regularly, as recommended by the manufacturers.
b. Avoid exposure to excessive humidity, heat or cold for prolonged periods.
c. Strip guide, reflectance disc, and optical window should be checked and cleaned daily.
d. The strips for use should be stored in a cool dark place at temperature less than 25°C.

Q. 10. What precautions should be taken before using different batch of glucostrips?

a. Firstly calibrate the glucose meter by entering the code found on the vial of the test strips or the chip that comes with the test strips into the glucose meter.
b. Inability to do this may lead to result inaccuracy up to 4 mmol/L.

Q. 11. What is the correlation between glucometer and laboratory value?

a. Glucose levels in plasma are generally 10%–15% higher than glucose measurements in whole blood (and even more after eating).
b. Blood glucose meters measure the glucose in whole blood while most lab tests measure the glucose in plasma.

Hence, laboratory values are higher than measured by glucometer.

Q. 12. What is newer technology to measure blood glucose?

Research is being done on non-invasive methods for measuring blood glucose, such as using infrared or near-infrared light, electric currents, and ultrasound.

a. *The glucowatch G2 biographer*: Designed to be worn on the wrist and uses electric fields to draw out body fluid for testing. The device does not replace conventional blood glucose monitoring. One limitation is that the glucowatch is not able to cope with perspiration at the measurement site. Sweat must be allowed to dry before measurement can resume. Due to these limitations and others, the product is no longer on the market.
b. *Spectroscopic measurement methods,* in the field of near-infrared (NIR), by extracorporal measuring devices, failed so far because at this time, the devices measure tissue sugar in body tissues and not the blood sugar in blood fluid. To determine blood glucose, the measuring beam of infrared light, for example, has to penetrate the tissue for measurement of blood glucose.

Q. 13. What additional features does the free style optium blood glucose meter have?

- No chip is required.
- Foil wrapped strips that protect against environmental factors and are convenient to store and carry.
- Starts test only when enough blood has been applied to minimize errors and reduce strip wastage.
- A second drop of blood can be applied within 5 seconds if initial drop size is too small, to reduce strip wastage.
- Ability to test for both blood glucose and blood ketones with strips.
- Small blood sample required—0.6 microliters
- Test time 5 seconds for quick results.
- Backlight to assist in low light testing. It has a large backlit display.
- It has a large memory with averaging.
- Free style auto-assist compatible—download up to 450 results and print or email reports.

Chapter 24

Clinical Refractometer

This is a hand-held optical instrument used for the estimation of specific gravity of urine and protein content of serum or plasma. In the neonatal unit, its use is primarily to assess hydration of the babies and plan fluid therapy based on the urine specific gravity.

The clinical refractometer has been designed for simple, rapid microanalysis in medical and paramedical fields. Scales are calibrated in terms of protein concentration of plasma or serum (grams/100 ml) and specific gravity of urine and refractive index difference. Determinations are precise and rapid. It requires a drop of fluid sample. One simply reads the value on the appropriate scale as seen through the eyepiece where the sharp boundary between dark and light fields crosses the scale.

Principle of the Analysis

A refractometer is a simple instrument used for measuring solute concentrations of aqueous solutions (Fig. 24.1).

When light enters a liquid it changes direction; this is called refraction. Refractometers measure the degree to which the light changes direction, called the angle of refraction. A refractometer takes the refraction angles and correlates them to refractive index (nD) values that have been established. Using these values, we can determine the concentrations of solutions. For example, solutions have different refractive indexes depending on their concentration in water.

The prism in the refractometer has a greater refractive index than the solution. Measurements are read at the point where the prism

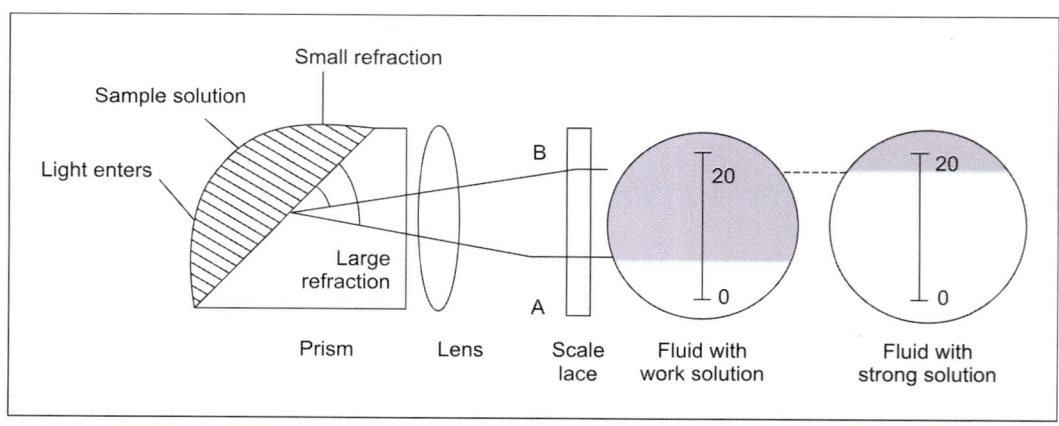

Fig. 24.1: Principle of refractometer

and solution meet. With a low concentration solution, the refractive index of the prism is much greater than that of the sample, creating a large refraction angle and a low reading ("A" on diagram). The reverse would happen with a high concentration solution ("B" on diagram).

This principle has been used to measure concentration of total solids or water in plasma and urine.

Its clinical application has been found in measuring the specific gravity of urine. It can also be used in estimating total protein in serum; as refractive index of serum or plasma depends mainly on its protein concentration.

The concentration of water in serum (grams/100 gm) and percent water in serum (grams/100 dl) can be determined from conversion table provided.

Urine specific gravity depends on the solute load. Where human urine is hypogravic (specific gravity less than 1.017) under concentration test, estimation of specific gravity by refractometric means is exceptionally accurate, regardless of variation in relative composition (e.g. salt, urea). However, with values in excess of 1.035, refraction correlates poorly with specific gravity. In these situations, total solids should be used for measuring concentration. Total protein in urine can be measured refractometrically by the determination of total solids in the fluid before and after protein has been removed by heat.

The instruments are temperature compensated for temperature between 60°F and 100°F, so that the reading need not be adjusted for either the temperature of the sample or the temperature of the room in which instrument is kept.

TYPES OF REFRACTOMETER

Different types of clinical refractometers are available in the market. They are the bench top refractometers, hand-held refractometers and the digital refractometers (Figs 24.2, 24.3 and Table 24.1).

The bench top apparatus can be used to read the SG of urine and the total serum protein concentration, by using the different scales given on the same machine. The handheld

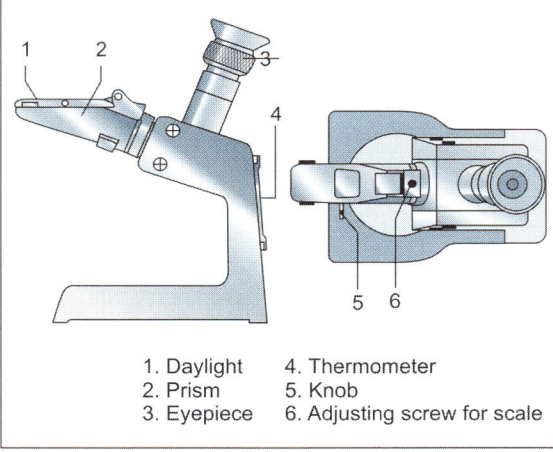

1. Daylight
2. Prism
3. Eyepiece
4. Thermometer
5. Knob
6. Adjusting screw for scale

Fig. 24.2: Clinical refractometer

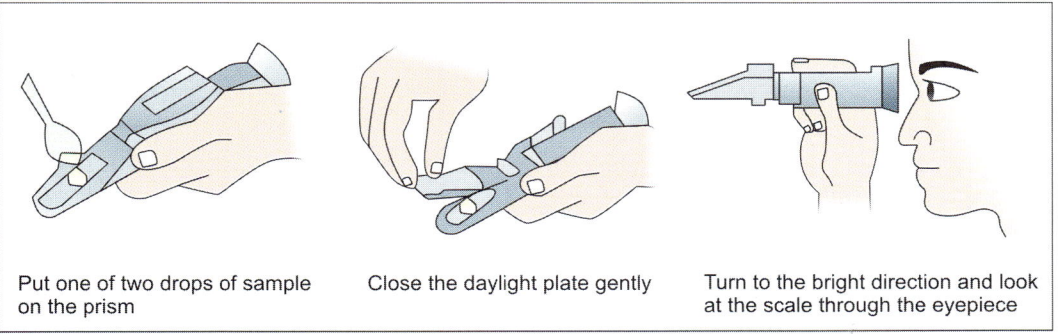

Put one of two drops of sample on the prism

Close the daylight plate gently

Turn to the bright direction and look at the scale through the eyepiece

Fig 24.3: Method of measuring through hand-held refractometer

TABLE 24.1: Common brands of clinical refractometer available in Indian market

S.No.	Make	Principals	Dealers	Unit cost (₹)
1.	AO Reichert TS meter 1310400A	AO Reichert scientific instruments	Nu-Tek Instruments	11,000
2.	Clinical refractometer T2-NE	Atago, USA Sper scientific direct	—	8,000
3.	Clinical refractometer Model SU-202	Optima	Mediserve	30,000
4.	Digital hand-held urine specific gravity refractometer K02941-70	Atago, USA	—	—
5.	Clinical refractometer EW-02942-05	Atago, USA	Cole-Parmer (India) Godrej Coliseum 101A Somiya Hospital Road, Mumbai 22	30,000
6.	Digital clinical refractometer	—	ACMAS Technolgy Pvt. Ltd.	50,000

refarctometers come in two different models (incorporating different scales) for measurement of urine SG or the serum proteins. Newer hand-held models that can be simultaneously used for both purposes, are also available. The digital one uses microprocessors and are battery operated. They display digital results instantaneously.

Scale Adjustment

Before measuring the specific gravity or serum protein concentration, the instrument should be checked as follows:

- Raise the daylight plate and place a few drops of distilled water on the face of the prism. Close the cover plate gently but firmly. This spreads a minimal volume of sample in a thin even layer over the prism.
- Bring the scale into focus by turning the eyepiece. If the boundary line does not coincide with waterline or specific gravity of 1.000, make an adjustment by rotating the scale adjusting screw-by-screw driver. This setting need not be changed as long as the same individual continues to use the instrument.

[Some instruments have a cement seal over the adjusting screw through which the screwdriver needs to be pushed. In these, the hole needs to be scaled with caulking compound after correct reading has been obtained (caulking compound is supplied with the instrument)].

Measuring Method

- Open the daylight plate and place one or two drops of the sample on the prism surface. Close the daylight plate gently so that the plate comes into contact with the prism.
- Point the end of the daylight plate in the direction of a bright light, rotate the eyepiece while looking through it until the image is correctly adjusted and the scale becomes clearly visible.
- A boundary line that separates the brighter and darker sides at the upper and lower portions respectively appears in the fields of vision. The reading is shown by this boundary line and indicates the protein concentration or urine specific gravity.

Precautions

1. The refractometer is an optical instrument, do not drop it or handle it roughly.
2. The prism has a relatively soft surface, be careful not to scratch it.
3. Use soft cloth or soft tissue moistened with water for wiping the prism. Dry the prism with a soft cloth or tissue. If the prism surface or cover plate is not well cleaned

before the next sample is loaded, an erroneous or fuzzy reading may result.
4. If the prism surface is smeared with oil or similar liquids, it will repel the sample and obstruct the measurement resulting in erroneous or fuzzy reading. Wipe off the oil smear with weakened detergent.
5. Do not immerse the eyepiece or the black focussing ring in water.

Frequently Asked Questions (FAQs)

Q. 1. Can the clinical refractometer be used for concentration of other body fluids?
Fluids such as pancreatic juice, saliva, etc. may also be analyzed refractometrically. However, interpretation of such refractions should not be made in terms of total solids or components thereof without reference to suitable standardization; it may be inappropriate to use the serum scale directly for this purpose.

Q. 2. How does one calibrate clinical refractometer?
To calibrate the instrument raise the daylight plate and place a few drops of distilled water on the face of prism. Close the plate gently. Now bring the scale into focus by turning the eyepiece. The boundary line which separates the brighter and the darker sides at the upper and lower portions respectively should coincide with the waterline or specific gravity of 1.000. If it does not, make an adjustment by rotating the scale adjusting screw-by-screw driver. Turn clockwise to increase reading, counterclockwise to decrease reading. Make sure that final motion is clockwise. Seal hole with caulking compound after correct reading has been obtained.

Q. 3. Give tips for interpretation of results for serum or plasma protein concentration.
Place a drop of plasma (from a pre-centrifuged heparinized capillary tube) on the refractometer. Record baseline (i.e. on day one of life) 'total solids' on the refractometer. Hence, onwards, further values can be compared with baseline value for that patient—giving an idea about hydration status.

Q. 4. What is normal specific gravity of urine? What does one infer from low or high urine SG report?
Normal SG of urine in the neonate varies from 1005–1015.
If SG >1015 (especially >1020), think of
- Concentrated urine
- Under hydration
- Hypovolemia
- Increased solute load (e.g. glucose, sucrose, protein, radiopaque iodine compounds, sodium sulfate, etc.)

If SG <1005
- Excess fluids
- Diuretic phase of ARF
- Diabetes insipidus.

Q. 5. Give tips for use of clinical refractometer.
i. Do not handle roughly or drop as this is an optical instrument.
ii. Be careful so as not to scratch the prism area.
iii. Keep instrument dry and clean.
iv. Check calibration periodically.
v. Never expose the instrument to temperature above 150°C.

Q. 6. What are the characteristics of new instantaneously measuring clinical refractometer?
New clinical refractometer powered with rechargeable Ni-Cd battery measure specific gravity in urine and total albumin in serum within one second. The sample has to be put on the sample stage and the instrument turned on. Reading is depicted instantaneously.

Q. 7. What is automatic temperature compensation (ATC)?
ATC allows user to take accurate scale readings at varying ambient room temperatures. The benefit include that with variable ambient temperature by a degree centigrade or two, the refractive index may shift. ATC refractometer require less calibration, maintain accuracy at wider range of temperatures (20–50°C).

Chapter 25

Spectrometric Bilirubin Analyzer (Micro-method for Serum Bilirubin Estimation)

Neonatal jaundice is one of the most common problems encountered, during the care of a newborn baby. The accurate and fast, bedside estimation of bilirubin is important in clinical management of jaundice in these babies. It is tempting to use the intensity of the colour of bilirubin for determining its concentration in serum. This temptation is all the more compelling while managing newborn infants, because newborns require frequent sampling and quick therapeutic decisions. Hence, we need a simple, reliable, accurate, laboratory test which is rapid and uses minimal amount of blood. The method of direct spectrophotometry, which is used in the modern day bilirubinometer, seems to satisfy these requirements.

PRINCIPLE OF SPECTROPHOTOMETRY

The concept of spectrophotometric estimation of serum bilirubin, is based on the Beer-Lambert's law. Whenever white light passes through a translucent solution of a given solute, some wavelengths are absorbed and the remaining passes through. The intensity of the light, which has passed through, can be measured at different wavelengths. From this an absorption curve can be constructed that tells us the wavelength interval in which the solute is absorbing light and the wavelength at which there is peak absorption. Once the absorption curve for a given solute is known, it is possible to pass light at the required wavelength and detect how much of the light is absorbed. The concentration of the solute can then be calculated. To summarize the Beer-Lambert's law—concentration of solute in a solution can be determined by the following formula:

$$L (out) = L (in) - (DCa)$$

where,

L = Intensity of light

C = Concentration of solute in solution

D = Distance through which light travels

a = Absorption coefficient of solute.

SPECTROPHOTOMETRY ESTIMATION OF BILIRUBIN

The situation is not so simple when we are measuring a particular substance in serum, because serum has a complex mixture of substances. For example, in the case of bilirubin, there are five other serum components whose absorption curves overlap in the same wavelength interval, viz. carotenoids, oxyhemoglobin, transferrin, *methemalbumin* and *turbidity components*. The absorption curves of these components are known and using complex mathematical formula correction factors have been devised. However, the problem still remains with respect of carotenoids and oxyhemoglobin because their absorption peaks are very close to that of bilirubin. *The interference by oxyhemoglobin* has been circumvented by exploiting the fact that oxyhemoglobin has secondary absorption peaks at higher wavelengths. Hence, if light

Spectrometric Bilirubin Analyzer (Micro-method for Serum Bilirubin Estimation)

at two wavelengths is passed, namely at 454 nm (absorbed by bilirubin, oxyhemoglobin and carotenoids) and at 540 nm (absorbed only by oxyhemoglobin) the interference produced by oxyhemoglobin can be calculated and subtracted. The level of beta-carotene being negligible in the newborn period, direct spectrometry can be used in the newborn period although it becomes unreliable thereafter.

How does the bilirubinometer actually work?

As the bilirubinometer works on the principle of two wavelength direct spectrometry, it employs a light source which emits a narrow beam of light at 454 and 540 nm. This beam passes through a slit in the microcapillary tube holder or a couvette and the unabsorbed light is detected by a photodetector.

The microcapillary tube containing patient's blood is blocked on one end and centrifuged at 12,000 rpm for 5 min, to separate out the plasma or serum. In case of couvette, the serum or plasma is taken. It is then fixed onto the holder ensuring that the plasma or serum column covers the entire length of the slit through which the light passes. A microprocessor converts the lights intensity received by the photodetector into the total bilirubin value, which is then digitally displayed. There is no difference between plasma and serum samples as far as bilirubin values are concerned (Fig. 25.1).

Advantages of Spectrometry Over the Standard Diazo Method

1. The spectrometer is a small instrument and can be kept inside the nursery or in an adjoining side laboratory enabling quick bedside estimation.
2. The method is simple and generally requires no reagents. Some authors have used caffeine or borate to dilute the sample and improve the accuracy.
3. Bilirubin can be estimated from a microcapillary sample requiring only very small volume of blood (50–70 microliters) and the results are instantaneous.
4. The hematocrit can be read off the same sample, and the serum can be subsequently used for C-reactive protein determination, thus minimizing blood sampling.
5. Bilirubin values are not altered by creatinine.
6. Spectrometry is reliable even when there is hemolysis, unlike Doumas's diazo method in which azobilirubins get destroyed when HbO_2 levels exceed 50 µmol/L.

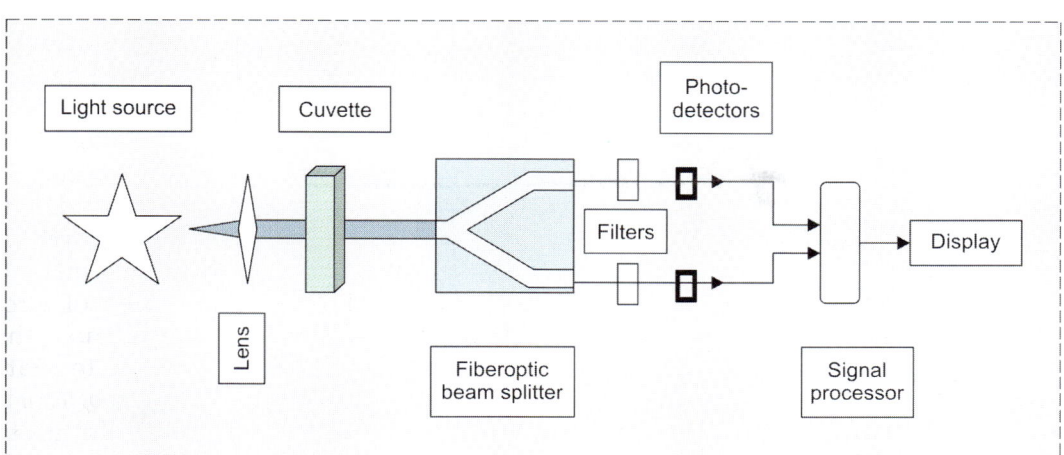

Fig. 25.1: Pictorial representation of the functioning of direct dual wavelength spectrophotometer used in the bilirubin analysis

Limitations of Spectrometry

1. Since various correction factors are applied for the interfering serum components there is an inherent inaccuracy. The gold standard for measuring bilirubin is high pressure liquid chromatography (HPLC), which unfortunately is very cumbersome.
2. Use is limited to newborn period because beta-carotene levels rise thereafter and interfere with results.
3. Conjugated fraction cannot be separately determined.
4. Cannot be used for newborns receiving intravenous lipids because lipidemia markedly increases turbidity of the serum/plasma.

Useful Tips

a. *The bilirubinometer reading in a baby and the clinical assessment of level of jaundice are not compatible*
   1. Make accurate clinical assessment. Clinical assessment has its own fallacies. Get the baby reassessed in daylight. Make sure the skin is not bleached or bronzed by phototherapy.
   2. Make sure dust has not accumulated in the bilirubinometer. It may interfere with light generation and detection. Particles of plasticine which are used to block the end of the capillary tube may also stick to the insides of bilirubinometer.
   3. Ensure that serum or plasma column covers the entire length of slit. If it does not cover the slit entirely, too much light will pass; and if the packed red cell column partially covers it, too little light will pass.
   4. Ensure that the blood sample has not been left exposed to light as this may cause photodegradation of bilirubin.
   5. Check whether the sample has been well centrifuged otherwise suspended red cells in the plasma column will interfere with accuracy of estimation.
   6. Recalibrate the bilirubinometer using blank and the standard solutions.
   7. Send a laboratory sample to countercheck.

b. *The laboratory value and the bilirubinometer readings do not show correlation*
   1. Go through the checklist as in the previous section.
   2. Recalibrate the bilirubinometer.
   3. Talk to the laboratory personnel. The diazo methods are also prone to problems of standardization and bilirubin estimation may have interlaboratory variations.
   4. Make sure that one is not dealing with a clinical situation where discrepancies are expected—renal failure with high creatinine, post-exchange transfusion, post-neonatal period, infant on intravenous lipids and so on.
   5. Reassess the patient and correlate clinically.

Specifications for an Ideal Spectrophotometric Bilirubin analyzer (Figs 25.2 and 25.3)

1. Small compatible, bench top, point of care bilimeter
2. Direct photometry determination of TSB with display in mg/dl or µmol/L
3. Automatic calibration between measurements and compatibility with any microcapillary tube
4. Dual wavelength 460 +/– 10 nm and 550 +/– 10 nm with correction for hemoglobin at 550 nm

Fig. 25.2: Spectrophotometric twin beam bilirubin analyzer

Spectrometric Bilirubin Analyzer (Micro-method for Serum Bilirubin Estimation)

Fig. 25.3: Spectrophotometric single beam capillary bilirubin analyzer

5. Accuracy of at least, within 5% of the laboratory spectrophotometer assay
6. Analysis time ideally less than 5 seconds
7. Large LED display
8. Appropriate device safety and quality certification like CE certification and ISO certification.

Frequently Asked Questions (FAQs)

Q. 1. What is the ideal way to estimate serum bilirubin level in a neonate?

Bilirubin analyzer which works on the principle of two-wavelength direct spectrometry (454 and 540 nm) utilising microcapillary tube or a couvette requiring micro blood sample is ideal for estimating serum bilirubin in newborn. The instrument is small and can be kept inside the nursery. The method for estimation of bilirubin is simple, requires no reagents and needs only very little blood sample (50–70 µl).

Q. 2. What are the characteristics of a good bilirubinometer?

1. It should be easy to calibrate.
2. The instrument should not be bulky, it should be easy to use and recalibrate.
3. It should use two-wavelength spectrometry. Instruments using multiple wavelengths for each interfering serum component have not shown any advantage over two-wavelength models.
4. It should be tropicalized for use in Indian climates.

Q. 3. How is the bilirubinometer calibrated?

The absorbance scale is checked by using standard potassium chromate solutions or the US national bureau of standards neutral glass filters. The wavelength calibration is done using standard commercial filters such as Spectest and Jorgen Fog. The calibration solutions should either be provided along with the instrument or should be readily available from the manufacturers. Ideally, the instrument should be calibrated daily.

Q. 4. Is transcutaneous bilirubin analyzer a good alternative?

Transcutaneous bilirubin analyzer is used as a screening tool (Table 25.1). The abnormal values detected will have to be cross-checked by bilirubin estimation. Transcutaneous bilirubinometry works on the principle of reflectance spectrophotometry and may give erroneous readings due to dark/pigmented skin. Latest Bili-chek system claims that the correlation between transcutaneous readings and serum levels of bilirubin is good, irrespective of the color of the skin and the gestational age.

Q. 5. What are the possible reasons for discrepancy in bilirubin estimation by bilirubinometer and clinical assessment?

Clinical assessment of neonatal jaundice has its own fallacies. Baby should be reassessed in daylight and remember that one should not comment on bilirubin level in a baby who is exposed to phototherapy. The bilirubinometer may not be functioning optimally. Very

TABLE 25.1: Common available brands of imported bilirubin analyzers in Indian market

S.No.	Make	Principals	Dealers	Unit cost (₹)
1.	One beam	Ginevri, Italy	Global Medical System	3,50,000/-
2.	Twin beam	Ginevri, Italy	Global Medical System	3,50,000/-
3.	Bil: Micrometer	Kohsoku Denki Co. Ltd.	Medi Equip India Pvt. Ltd.	1,30,000/-
4.	American Opticals—couvette method	American Opticals	—	1,50,000/-
5.	Cosmo Microcapillary method	Cosmo Medical	Medical Systems, Global Medical System	1,10,000/-
6.	Tot: Bil plus	DAS, Italy	Mediserve	1,50,000/-
7.	BR 5100: Dual wavelength Total bilirubinometer	Apel Co. Ltd., Japan	Global Medical Systems	1,50,000/-
8.	BR2 bilirubin stat analyzer Direct and total bilirubin	Advanced Instruments, USA	Global Medical Systems	4,50,000/-
9.	Single beam	DAS SRL, Italy	M/s Cardiocare Mobile: 09417005769	2,25,000/-
10.	Bil-100 bilirubin analyzer	Cosmo Meditech, South Korea	Krish Biomedicas	3,00,000/-

often the problem is due to accumulation of dust or particles of plasticine (used for blocking the end of capillary) inside the bilirubinometer. One must ensure that serum or plasma column should cover the entire length of slit through which the light waves passes. Blood sample left exposed to light may cause photodegradation of bilirubin leading to under estimation.

Q. 6. What are the factors which give rise to erroneous bilirubin estimation while using direct spectrophotometric bilirubin analyzers?
The bliirubin estimation by bilirubinometer may be erroneous in situations of high creatinine, in post-exchange transfusion with adult blood due to higher concentration of carotene, in post-neonatal period again due to increasing carotene levels and in babies with lipemic serum especially when they are on total parenteral nutrition.

Q. 7. Is there an instrument to measure total and direct bilirubin?
Advanced bilirubin stat analyzer model BR2 is only instrument that measures total and direct bilirubin. Direct bilirubin is determined by the method of Evelyn–Malloy. Blood sample is acidified and diazo reagent is added. A color reaction with the direct bilirubin is induced and direct bilirubin concentration is measured within 2 minutes.

Chapter 26

Laboratory Microcentrifuge

When a solid particle suspended in a liquid medium is allowed to stand in a tube, the solid particles gradually settles down at the bottom of the tube by the effect of gravity. However, this takes a long time. This process of sedimentation can be accelerated when a force greater than gravity, such as *centrifugal force* is used. This process of separation of solid from liquid is called *centrifugation* and the equipment used for this purpose is called *centrifuge*. A centrifuge is a piece of equipment that puts an object in rotation around a fixed axis (spins it in a circle), applying a strong force perpendicular to the axis (outwards).

Principle of centrifugation: The centrifuge works using the sedimentation principle. Let us consider the example of blood. Blood contains cells suspended in plasma. When it is rotated at a high speed in a centrifuge, a centrifugal force is generated which moves heavier and denser cells away from the center. Plasma gets separated and collects in the part of the tube closer to center. The centrifugal force generated can be calculated by the following formula:

$$C_f = 1.118 \times 10^{-6} \times R \times N^2$$

Where C_f is centrifugal force, R is rotational radius (center of axis to the tip of the tube) and N is speed of rotation (in revolutions per minute or rpm).

Following are the factors affecting centrifugation:
1. The speed of revolution
2. Duration of application of centrifugal force
3. Particulate matter concentration, i.e. packed cell volume. Higher the PCV more is the force needed for separation.

Factors affecting ESR such as viscosity of the blood or plasma protein concentration do not affect as greater centrifugal forces negates their effect.

TYPES OF CENTRIFUGE

1. **Hand-held centrifuge:** Here centrifuge has a handle, which is manually rotated. It generates low speed only.
2. **Motor driven centrifuge:** Here, a motor powered by electricity is used for rotating tubes. It can be of two types:
   a. *Angle headed*—here tubes are held fixed at an angle of 30°–50°. They can rotate at a higher speed but sediment is formed at sidewall of the tube (Figs 26.1 and 26.2).
   b. *Horizontal headed*—here, tubes are attached to central axis by a trunion. Tubes are placed vertically and when they are rotated, they move outwards and assume horizontal position.
3. **Microcentrifuge**—they are specially used for packed cell volume determination of blood. They accommodate capillaries and can rotate at a high speed of 10,000–15,000 rpm (Figs 26.3 and 26.4). Common brands available in Indian market are shown in Table 26.1.

Other types such as supercentrifuge and ultracentrifuge are used in specialized laboratories.

160 Neonatal Equipment

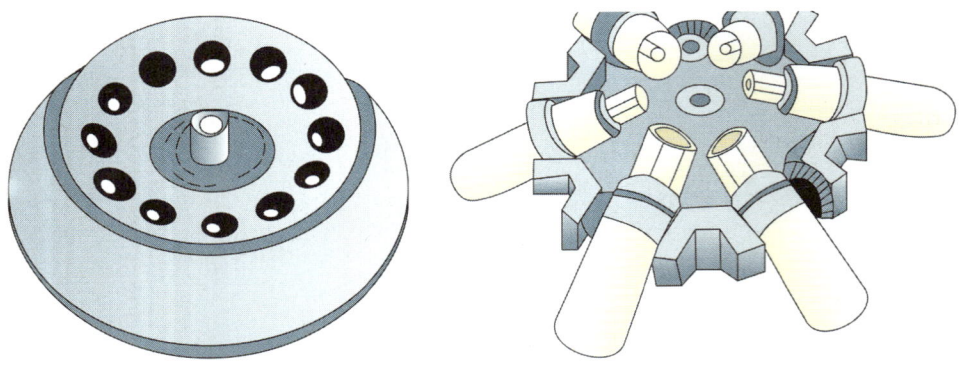

Figs 26.1 and 26.2: Angle head rotor (top); horizontal rotor (bottom)

Fig. 26.3: Microcentrifuge (A); Top view with closed lid (B); Top view with open lid shows space to stack filled capillary tubes (C)

Fig. 26.4: Microcentrifuge

APPLICATIONS OF THE CENTRIFUGE IN NEONATOLOGY

1. Measuring hematocrit in a microcapillary sample.
2. Separation of plasma from RBC for measurement of bilirubin in a capillary sample.
3. Centrifugation of urine for examination of sediment.
4. Centrifugation of CSF for examination for bacteria.

Microcentrifuge is used for 1 and 2.

MICROCENTRIFUGE

A microcentrifuge contains a solid-wall cylinder rotating about a horizontal or vertical axis. An annular layer of liquid is held against the wall by centrifugal force. Heavy phases move outwardly from the center, and less dense phases remain inwardly.

They accommodate microcapillary tubes and rotate them at a high speed of 10000–15000 rpm. The capillary tubes used for this purpose is made of borosilicate glass. Only microcapillaries certified for centrifugation at this

Laboratory Microcentrifuge

TABLE 26.1: Common microcentrifuge available in India

Model	Manufacturers	Dealers	Price
Microcentrifuge RM 12C	Remi Instruments Ltd.	Embee Diagnostics/ Ved Medical	₹ 15,000
Microcentrifuge	Damon capillary micro-centrifuge	—	₹ 50,000
Microcentrifuge	Medline Scientific Ltd., UK	Super Biotech	₹ 2,10,000
Microcentrifuge 1-14K	Sigma Laborzentrifuen Gmb	Sigma SVI Biosolutions	₹ 60,000
1-14K	BOECO Distributors	Nu Life	₹ 80,000
Microcentrifuge	Digital hematocrit centrifuge	Radical Scientific Equipment Pvt. Ltd.	

speed should be used. These capillaries have an average dimension of 1 mm of internal diameter by 7 cm of length. Heparin is the anticoagulant used in this method.

Parts of microcentrifuge: The external components of microcentrifuge includes a rotor with slots for capillaries, a cover disc with screw, a lid with lid lock, a timer with timer switch for setting time (for setting duration of centrifugation) and a knob for adjusting speed. Some important internal components include centrifuge motor, a belt drive and centrifuge motor brush.

CARE OF MICROCENTRIFUGE

1. Only recommended capillaries should be used.
2. For every tube placed in a slot, another tube should be placed in opposite slot for balancing.
3. Instrument should be kept clean and spill over of blood should be cleaned promptly.
4. Proper lubrication of the equipment should be done and motor brush should be inspected once in 3–4 months.
5. Instrument should be placed at least 30 cm away from wall for proper heat dissipation (during rotation heat is generated and needs to be dissipated. Most of present microcentrifuges uses continuous airflow through the centrifuge housing to restrict the temperature rise of the samples to a standard value of 40°C).

Frequently Asked Questions (FAQs)

Q. 1. How to measure hematocrit using microcentrifuge?

First of all, blood sample has to be collected in a pre-heparinised capillary tube. This sample can be either capillary blood or venous blood sample. The one end of capillary is sealed using plasticine.

Centrifugation of the sample: Following steps are followed for centrifugation of the sample:
a. Open the lid of the microcentrifuge and unscrew the cover.
b. Place the capillary in one of the slot of micro centrifuge.
c. Place another sample or a capillary filled with equal amount of water in the opposite slot.
d. The cover disc and lid are closed.
e. The centrifuge is turned on and speed is set to 10000 rpm for duration of 4–5 min.
f. At the end of time rotor is allowed to slow down by itself and capillary can be taken out after rotor stops (some microcentrifuge may have an in-built brakes to slow down the rotor. However, no external force should be applied to slow down the rotor and it should be allowed to slow down by itself).

Measuring the hematocrit: As capillaries are nongraduated. PCV is read using a micro-hematocrit reader. Two different types of PCV reader are available. One type measures the

PCV while the capillary tubes are still in the microcentrifuge while the other type is completely separate from the microcentrifuge.

Method 1: Measuring the packed cell volume on the microcentrifuge

Centrifuge the capillary tube as described above. Then place the Perspex reader over the plate holding the capillary tubes. While holding the plate still with one hand so that it does not turn, twist the knob on the reader with the other hand until the baseline (i.e. 0) crosses the capillary tube at the point where the red cells meet the plasticine. Now hold the knob still with one hand and rotate the Perspex reader with the other hand until the top line (i.e. 100%) crosses the capillary tube at the top of the serum (not the top of the tube). Determine which line crosses the capillary tube at the point where the red cells meet the serum. Follow that line along to either the left or the right and read the PCV.

Method 2: Measuring the packed cell volume off the microcentrifuge

Remove the capillary tube from the microcentrifuge and place it in the vertical groove of the reader so that the junction of the plasticine and the red cells lies on the bottom line. Slide the capillary tube holder to the left or the right until the top of the serum (not the top of the tube) falls on the top line. Move the Perspex arm up so that the line falls on the junction of the red cells and the serum. Now, read the PCV (Fig. 26.5).

Q. 2. What is the speed to be used for sediment examination of urine?

For examination of sediment in urine, sample is rotated in centrifuge in a test tube at a speed of 3000 rpm for 5 mins. The supernatant is removed using a pipette and sediment is transferred to a slide for examination.

Q. 3. How to minimize spilling of sample while rotating?

Spilling occurs when capillary is rotated at high speed and is subjected to undue shaking.

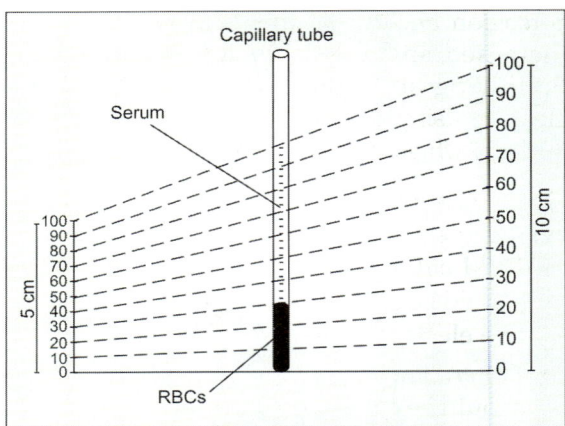

Fig. 26.5: Microhematocrit reader with centrifuged capillary tube hematocrit 30%

Use of appropriate size capillary fitting in slot and using capillary made of only recommended material minimizes chance of spilling. Also balancing the capillary by placing another capillary in opposite slot in rotor is essential.

Q. 4. What are the specifications to be looked for when buying a microcentrifuge?

A microcentrifuge must have:
1. Protection from overheating
2. Dynamic brakes
3. Rotation up to 16000 rpm, adjustable in increments of 100
4. Timer settable in minutes
5. Safety lid lock feature
6. Imbalance detector with cut off
7. Digital display shows rpm and time
8. Hematocrit reader (optional)
9. Low noise level

Do ask for the warranty of carbon brushes and extra carbon brushes.

Q. 5. How to clean the microcentrifuge?

The purpose of cleaning is, beside hygienic reasons, the avoidance of corrosion by soiling. Routine cleaning can be done with plain water. After cleaning ensure that all parts are dried thoroughly. All anodized aluminum parts should be regularly treated with anticorrosion oil, so that their durability will be increased and the corrosion risk reduced. Following spilling of blood, it should be cleaned using a neutral disinfectant such as 10% bleach solution.

Q. 6. How to verify functioning of microcentrifuge?

i. *RPM calibration*: The speed of centrifuge should be checked by using external device like electronic tachometer.

ii. *Timer*: The timing intervals of the timer should be checked against an accurate stop watch or electronic timer.

Q. 7. What preventive maintenance steps should be ensured for optimal functioning?

i. For preventive maintenance such as lubrication of bearings, major components and replacement of graphite brushes, current leaking and checking of fuse should be done according to the manufacturer's direction.

ii. The line cord, plug and control knobs should be examined on regular intervals.

iii. The rotor, lid and control knobs should be examined frequently for mechanical stress (cracks and corrosion).

Chapter 27

Laminar Airflow System

The control of microbial and particulate contamination has become increasingly important in areas catering to patients whose normal defense mechanisms are impaired. This applies especially to a neonatal intensive care unit (NICU) with immunologically immature preterm babies. A practical solution to the control of microbial contamination in the NICU is the use of laminar airflow system.

AIRBORNE CONTAMINATION

Air is the vehicle for the transport of small particulate matter including micro-organisms. Air suspended particles deposit themselves on surfaces thereby contaminating objects and also deposit in the respiratory tract of human. Since uncontrolled airborne contamination is detrimental to the success of any intensive care unit, they should have adequate measures to control the same. These measures paved way for the concept called 'clean room technology'.

Clean room technology
This works on the basis of the following steps:
1. *Filtration of air*: By using filters that remove all airborne particulates, e.g. high efficiency particulate air filter (HEPA filter).
2. *Distribution of filtered air by laminar flow:* This sweeps away any internally generated material rapidly out of the environment.

The laminar flow increases the efficiency by 4 to 5 times the level that could be achieved with only filtration of air. Thus, it was implemented in the 'clean rooms' since late 1950s.

LAMINAR AIRFLOW SYSTEM

A. Principle of Laminar Airflow

Laminar flow is the movement of air which is continuous, unidirectional, and with a uniform low velocity in one direction along parallel flow lines, either horizontally or vertically. This flow, being devoid of turbulence and backflow, creates an environment with extremely low levels of contamination. Any contaminated particles will not be picked up from one spot and deposited in another (as in turbulent flow). Rather, it will be carried away from the working area to the exhaust point (Fig. 27.1).

B. Components of Laminar Flow System

These systems contain 3 basic elements:
- HEPA filter
- Blower
- Plenum

High Efficiency Particulate Air (HEPA) Filter

HEPA filters are designed to remove particles, including micro-organisms from the air. Though very effective at trapping particulates and infectious agents, they are not able to remove volatile chemicals or gas. These filters are made of boron silicate micro-fibers that are made into a flat sheet by a process similar to paper making. The flat sheets are pleated to increase the overall surface area of the filter. Pleats are separated by aluminum baffles that direct the airflow in the filter (Fig. 27.2).

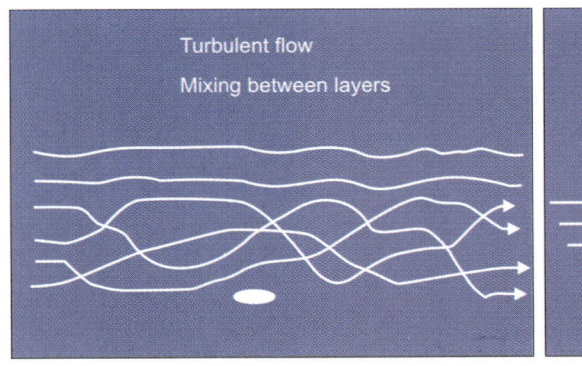
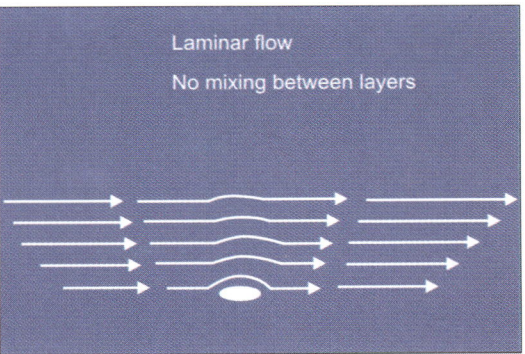

Fig. 27.1: Turbulent and laminar airflow

HEPA filters

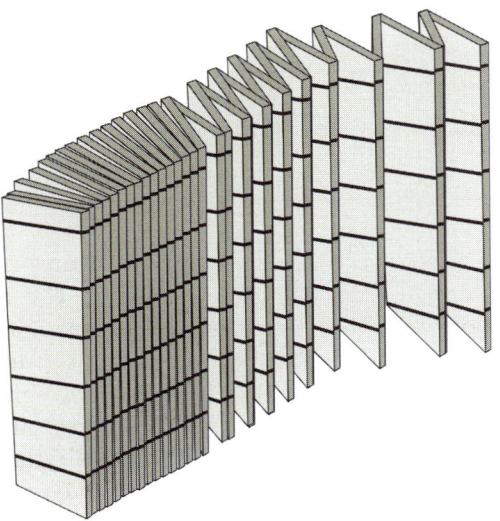

Fig. 27.2: Structure of HEPA filters showing the pleated structure of boron micro-fibers

HEPA filter acts by one of three mechanisms: Diffusion, interception or impaction. It can reliably remove 99.97% of all the particles of 0.3 m size (i.e. 9,997 of every 10,000 particles). 0.3 m size is known as the most penetrating particle size (MPPS). Particles that are *smaller* or *larger* than 0.3 m size are trapped with even *higher efficiency*. Since the critical particle diameters for deep lung deposition are in the range of 1 to 5 m and also the size of most of the bacteria being >0.3 m, the actual efficiency of HEPA filters is more than 99.97%.

C. Mechanism by which a Laminar Airflow System Works

Laminar airflow system is illustrated in Fig. 27.3. The contaminated room air that enters the unit is initially filtered with a pre-filtration system to remove the larger dust particles. This pre-filter is made of polyurethane and can be cleaned with air vacuum. This pre-filter increases the efficiency of the HEPA filter. It needs no replacement over the operating life of the laminar flow, thus decreasing the maintenance cost. The air is then driven by a motor or blower through the HEPA filter. Clean air vacates the unit through the exhaust vent in a unidirectional way along parallel flow lines. Unidirectional airflow showers the work zone with a continuous supply of filtered air. This shower effect serves to sweep the contaminants out of the environment through the air exhaust system. The speed of

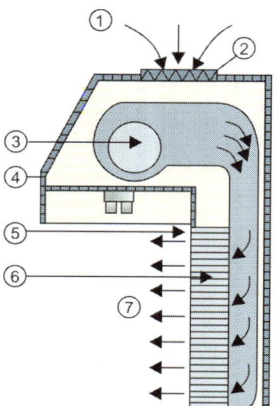

1. Dirty room air
2. Pre-filtration system
3. Motor/blower assembly
4. Negative pressure section
5. HEPA filter
6. Downstream gasket seal
7. Positive pressure laminar flow of clean air

Fig. 27.3: Basic principles of working of a laminar airflow (horizontal mode)

contamination removal is very high since the filtered air moves through the system in a single pass. In addition, the UV light (present in most of the laminar flow systems) keeps the area bacteria-free.

D. Classification of Laminar Airflow Systems

Laminar airflow devices can be classified into 3 major categories depending upon the direction of the airflow within the device/unit (Table 27.1). A vertical laminar flow has the additional advantage over the horizontal system in that particles are removed by gravity as well as by airflow. This is particularly important in the case of heavy particles that can fall out and settle.

For a neonatal ICU, only the clean rooms (with either horizontal or vertical airflow) and work hoods (work stations) are essential.

i. **Laminar airflow rooms:** This indeed forms the nucleus of 'clean room technology' (Fig. 27.4A). Though it is desirable for every NICU to have laminar airflow, there are certain practical difficulties in implementing it. The advantages and disadvantages of these rooms are discussed in Table 27.2.

ii. **Laminar flow hoods (work stations):** Ideally any fluid/drug preparation in a NICU including the preparation of IV fluids, drugs and total parenteral nutrition (TPN) solutions should be done using a laminar flow hood. Thus, it is essential for any unit treating sick preterm babies who require TPN infusions to have a laminar flow hood. A prototype horizontal airflow hood is illustrated in Fig. 27.4B.

TABLE 27.1: Examples of laminar airflow systems

Horizontal airflow units	Vertical airflow units	Curvilinear units
Rooms with wall to wall airflow	Rooms with grated floors	Rooms and solid walls
Work hoods	Portable units with curtains	Work hoods for walls
Walk-in booths	Work hoods	
	Biological safety cabinets	

TABLE 27.2: Advantages and disadvantages of laminar airflow rooms

Advantages	Disadvantages
1. Clean down capacity—the airflows carry away the contamination generated within the unit	1. High cost (especially with a vertical laminar flow system)
2. Improved control of humidity and temperature	2. Higher air velocities require larger fans, motors and ducts
3. Air particles have minimum contact with the protected items	3. Clearance required for vertical airflow rooms 4 to 6 ft of additional vertical height is needed
4. Reduced maintenance of the equipment	

Fig. 27.4A: Laminar airflow room (vertical)

Fig. 27.4B: Laminar flow hood (horizontal)

E. Maintenance

Laminar airflow hoods are designed to be operated continuously. If a laminar airflow hood is turned off between aseptic processing, it should be operated for 30 minutes to allow complete purging of room air from the critical area, and then disinfected before use. The critical area work surface and all accessible interior surfaces of the hood should be disinfected with either bacillocid or 70% isopropyl alcohol using sterile gauze at the beginning of each shift and periodically thereafter. The glass surfaces should be cleaned daily with soap/detergent only. HEPA filters should be tested every 6 months and pre-filters should be cleaned monthly.

Before usage (if not in continuous use), the front panel is closed and the UV light is switched on for 30 minutes (instead, the work surface can be disinfected with bacillocid also). After 30 mins, the UV light is switched off and the front panel opened. The fluorescent tube light and air blower are turned on. Airflow is adjusted according to the LCD display. After use, the panel is again disinfected with bacillocid and the front panel is closed.

Technical Specifications

The most common standard referred to by most manufacturers of laminar flow systems is the US Federal Standard 209E. It does not deal with laminar flow construction but with most important aspect of cabinet performance which is the level of product protection provided (i.e. cleanliness of air) within the working area of the laminar flow cabinet.

1. *Type of airflow*: Horizontal or vertical
2. *Size of working area*: For example, $4 \times 2 \times 2$ feet
3. *Work surface*: Finely polished stainless steel
4. *Body*: Preferably stainless steel construction
5. Foldable/sliding front sash, preferably of transparent plastic
6. Blower assembly producing a little noise/vibration
7. HEPA filters of international standards
8. Microprocessor based monitoring system with LCD display of airflow velocity, residual lifetime of HEPA filters and total time of cabinet operation
9. Alarms for clogged filters, out of range airflow velocity and other malfunctions
10. Fluorescent tubes for lighting
11. UV light for additional sterilization of the work chamber
12. Transparent UV tempered glass side walls
13. White non-reflective powder coated back wall

Common brands available in Indian market are shown in Table 27.3.

TABLE 27.3: Common laminar airflow systems available in Indian market

S.No	Manufacturers	Distributors	Unit cost (₹)
A. Imported			
1.	NuAire Inc, US	Bluestar Limited, Chennai	—
2.	Esco Micro Pvt. Ltd, Singapore www.biotech.escoglobal.com	Esco Biotech Pvt. Ltd., Mumbai	—
B. Indian			
1.	DM-188 Stericlean Horizontal airflow hood 4 × 2 × 2 ft 6 × 2 × 2 ft	Yorco Sales	3,00,000 4,00,000
2.	Horizontal airflow 4 × 2 × 2 ft	Sam Products	1,71,000
3.	Horizontal airflow 4 × 2 × 2 ft	JKG Bioscience	1,90,000
4.	SS, model, SB-LH2	Sunil Brothers	
5.	Hicon laminar airflow	Global Medical Systems/Grover Enterprises	
6.	Vertical laminar flow hood	MS Sunrise Enterprises	
7.	KS12/RN solution	Thermofisher Scientific	
8.	Bio11 advance 3	Mehrotra Biotech/Telstar Technologies	

(Other Indian manufacturers are Pheroh Filters and Equipment, Biotechnologies Inc [New Delhi]), Clean air systems, Chennai, Gautam Enterprises [Delhi], Sanlar Services [Mumbai].

Frequently Asked Questions (FAQs)

Q. 1. How should I decide whether laminar flow system is necessary for my unit?

a. *Laminar flow hood*: Any unit with sick term/preterm babies should preferably have a laminar flow work station. This is essential for sterile preparation of IV fluids, drug solutions and more importantly TPN fluid.

b. *Laminar airflow rooms*: This depends upon many factors—size of the rooms, height of the work space, capacity of the air-conditioning systems, the quality of the unfiltered outside air, and the cost. One should also remember that though they are effective, they have certain limitations (*see below*). If the facility is to be of the total room concept, the task of design, purchase and installation should be handled by an architectural/engineering group. Since such facility will not be feasible in most instances, mobile laminar units (similar in concept to incubators) can be a good alternative. This facility can be used for sick ELBW babies and babies with congenital/acquired immunodeficiency.

Q. 2. What is a laminar airflow system?
A laminar airflow system contains three basic elements—a blower, a high efficiency air filter, and a plenum. There may be variations on this idea—many blowers, many filters, and very large plenums, but all have the same basics. The flow is called laminar because the turbulent air upstream is changed by the filter into a straight-line flow off the downstream face of the filter.

Q. 3. What are the limitations of laminar airflow system?

a. It does not provide an air scrubbing action—so it would not remove particulate contamination already residing on the surfaces/objects.

b. It obviously cannot prevent the generation of contamination.

Thus, it is not a panacea for all the problems—again emphasizing the importance of good housekeeping practices, proper hand-washing and other aseptic techniques.

Q. 4. How are the laminar airflow systems evaluated?

They are evaluated by measuring airflow rates, turbulence patterns, traditional particle counting methods and microbiological sampling using modern day air and surface samplers.

The commonly used organisms for testing are aerosols of *E. coli* T1 bacteriophage and of *B. subtilis* var. *niger*.

Q. 5. Are there any international standards for the laminar airflow system?

The most common standard referred to by most manufacturers is the US Federal Standard 209E. It deals with the level of cleanliness of air within the working area of the laminar flow cabinet. There are 3 classes of air cleanliness based on particle counts. These classes are class 100, class 10,000 and class 100,000. In class 100, particle count does not exceed 100 particles per cubic feet of a size of 0.5 micron and larger. Some of the other standards used are BS 5295, ISO Standard 14664.

Q. 6. How are the high efficiency particulate air (HEPA) filters tested?

The HEPA filters are tested by the DOP (dioctylphthalate) method when manufactured. DOP, a liquid plasticizer is heated to the point of vaporization and reconstituted into 0.3 µ particles to form a mono-dispersed aerosol. These single size particles are diluted with air until a concentration of 100 µg/L is reached and then the aerosol-air mixture is passed through the filter. The amount of penetration is measured on the downstream side with a forward light scattering photo-meter, giving the familiar readings of 0.03% or better. The material used to make the filter is tested in the same way by the filter material manufacturer.

Q. 7. When should the HEPA filters be replaced?

The average HEPA filter, properly installed and with frequent changes of the pre-filter, should last from five to eight years. Otherwise, the resistance of the filter as indicated on a manometer or the airflow measured with a velometer is a good indicator of the need for a change.

Q. 8. What are the other applications of laminar airflow systems?

They can be used in a wide range of disciplines such as:

1. Quality control labs of micro circuit, electronic assembly and manufacturing applications.
2. Quality control labs of pharmaceutical and food processing industries.
3. Deoxyribonucleic acid: Thermocycling.
4. General laboratory applications in bio-technology and microbiology.
5. Handling of hazardous agents to human beings or animals as defined in the appropriate international standards.

Section G. Life-saving Equipment

28. Self-inflating Bag: Manual Resuscitator
29. Oxygen Concentrator
30. Continuous Positive Airway Pressure Machine
31. Heated Humidified High Flow Nasal Cannula Therapy
32. Neonatal Ventilators
33. Inhaled Nitric Oxide Delivery Systems

Chapter 28

Self-inflating Bag: Manual Resuscitator

The self-inflating bag, as its name implies, inflates automatically without a compressed gas source. It remains inflated at all times, ready for use. Since it is not dependent on a compressed source for inflation, it is portable. There are four parts of the self-inflating bag:

- Air inlet
- Oxygen inlet
- Patient outlet
- Valve assembly

AIR INLET

As the bag re-expands following compression, air is drawn into the bag through a one-way valve that may be located at either end of the bag, depending on its design. The opening for air is called air inlet.

OXYGEN INLET

Every inflating bag has an oxygen inlet, which is usually located near the air inlet. The oxygen inlet is a small nipple or projection to which oxygen tubing can be attached when oxygen is needed. In the self-inflating bag, an oxygen tube does not need to be attached in order for the bag to function. It has to be attached if the infant is to be resuscitated with an oxygen-enriched air mixture rather than with room air (Fig. 28.1).

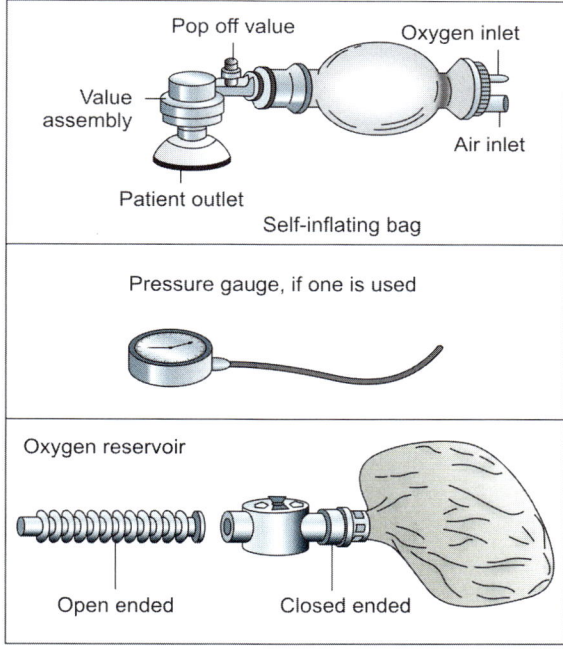

Fig. 28.1: Parts of self-inflating bag

PATIENT OUTLET

The patient outlet is where the air exits from the bag to the infant and is where the mask or endotracheal tube connector can be attached.

VALVE ASSEMBLY

Self-inflating bags have a valve assembly positioned between the bag and the patient outlet. When the bag is squeezed during ventilation, the valve opens, releasing oxygen/air to the lungs of patient. When the bag re-inflates (during the exhalation phase of the cycle), the valve is closed. This prevents the patient's exhaled air from entering the bag and being rebreathed.

Optional Parts

Pressure Gauge

Some self-inflating bags have a site for attaching a pressure manometer/gauge. The attachment site usually consists of a small hole or projection close to the patient outlet. A pressure gauge is an extra piece of equipment attached to the bag by means of a small tube, at a point close to the patient outlet. It measures pressure generated by the bag in centimeters of water. This gauge allows the person using the bag to control the pressure of the air or oxygen being delivered to the patient.

PEEP Valve

An adjustable PEEP (positive end expiratory pressure) valve can be connected to valve assembly. It is useful if one is resuscitating extremely low birth weight baby (<1 kg) or if manual resuscitation for a baby disconnected from the ventilator is required.

Using Bag with Oxygen

All babies requiring positive pressure ventilation at birth should be ventilated with a bag capable of providing higher concentration of oxygen (90–100%) but at delivery room use of blenders for regulating oxygen is recommended. In term babies room air resuscitation can be initiated with back up facility for supplementing oxygen if baby does not improve following 90 seconds of room air administration. In preterm babies <32 weeks resuscitation can be initiated with bag attached to oxygen source connected with blender and FiO_2 between 21–30%. Because oxygen is considered to be a drug and its use in neonates must be carefully controlled, it is important to know the approximate concentration of oxygen being administered to an infant during resuscitation.

Remember that oxygen can be brought into a self-inflating bag through tubing connected to an oxygen source. Each time the bag re-inflates, room air is drawn into the bag by way of the air inlet. This means that even though hundred percent oxygen is flowing through the O_2 inlet, it is diluted by the room air that enters each time the bag re-inflates.

As a result, the concentration of oxygen actually received by the patient is greatly reduced, it is somewhere in the range of 40–70%.

Oxygen Reservoir

High concentrations of oxygen can be achieved with a self-inflating bag through the use of both the oxygen tubing and an oxygen reservoir. An oxygen reservoir is an appliance that can be attached over the bag's air inlet. This reservoir provides a chamber filled with a high concentration of oxygen. During re-inflation, instead of room air being drawn in, the bag draws the highly oxygen-enriched air in the reservoir. This permits administration of as high as 90–100% oxygen with a self-inflating bag.

This means that self-inflating bags without oxygen reservoirs are totally inadequate for resuscitation in the delivery room. Therefore, all self-inflating bags used in a delivery room must have oxygen reservoirs attached so that the bags are capable of delivering a high concentration of oxygen.

Safety Features

Two safety features are built into resuscitation bags to help control the amount of pressure that goes into the lungs. These safety mechanisms prevent high pressures being delivered inadvertently to lungs.

These safety features are:
- The pressure release valve
- The pressure gauge.

Any resuscitation bag used for neonates should have *at least one* of these above two features.

A *pressure release valve*, more commonly known as a pop-off or safety valve, is a feature that is built into many resuscitation bags. These pressure release valves are set to release at 30–40 cm of water. Therefore, if pressures in excess of this limit are generated, the valve opens, preventing the excess pressure from being transmitted to the infant.

In some self-inflating bags, the pop-off valve can be temporarily occluded or bypassed to allow pressure in excess of 40 cm of water to be administered. This may occasionally be necessary to effectively ventilate a neonate's non-aerated lung, especially with the first few breaths. Extreme care must be taken not to use excessive pressure during the few ventilations in which the pop-off valve is bypassed. Any self-inflating bag in which one can bypass the pop-off valve should have a pressure gauge attached to it.

Use

1. To provide intermittent positive pressure ventilation.
2. To provide peak end expiratory pressure in preterm.
3. To judge pressure required before connecting baby to ventilator.

Misuse

1. Never use bag for providing free flow of oxygen.
2. Excessive pressures may result in pneumothorax.
3. Prolonged ventilation may lead to oxygen toxicity if 100% oxygen is being provided.
4. In thick meconium-stained neonate, do not use bag and mask until airway is cleared.
5. In diaphragmatic hernia, bag and mask ventilation will lead to acute deterioration. Use endotracheal tube and bag for resuscitation.

RESUSCITATION MASKS

Masks come in a variety of shapes, sizes and materials. The selection of a specific mask for use with a particular infant will depend on how well the mask fits the infant's face and how easy it is to use in obtaining a seal. Resuscitation masks have rims that are either cushioned or non-cushioned.

Non-cushioned: Some masks are constructed without a padded, soft rim. Such a mask usually has a very firm, abrupt edge to the rim.
- Because it does not easily conform to the shape of the baby's face, it requires greater pressure to form a seal than does a cushioned mask.
- It can damage the eyes if the mask is improperly positioned.
- It can bruise the neonate's face if the mask is applied too firmly.

Cushioned: The soft rim on a cushioned mask from either a soft, flexible material, such as foam rubber, or an air-inflated ring. A cushioned rim mask has several advantages over a mask without a cushioned rim:
- The rim conforms more easily to the shape of the infant's face, making it easier to form a seal.
- It requires less pressure on the infant's face to obtain a seal.
- There is less chance of damaging the infant's eyes, if the mask is incorrectly positioned.

Shape

Masks come in two shapes:
- Round
- Anatomically shaped.

Round: A round mask can be effective in obtaining a seal for ventilation. If the correct size is not selected, a seal cannot be formed, or it may not fit over the mouth and nose correctly. If the mask is too large, pressure may be exerted on the eyes and can cause damage.

Anatomically shaped: Some masks are shaped to fit the contours of the face. These masks are referred to as anatomically shaped masks. They are made to be placed on the face in a particular direction with the most pointed part of the mask fitting over the nose. It is easier to obtain a seal with an anatomical mask.

TABLE 28.1: Common resuscitation bags available in the market

S.No.	Type of bag	Dealer's name	Unit cost (₹)
1.	Silicone (indigenous)	Meditrin Mediserve, Zeal Medical, Phoenix Med, Delhi Hospital Supply	1000–1400
2.	Ambu bag (imported)	Indian Surgicals Equipment, Mediland Surgifield	3500–4500
3.	Silicone Laerdal Make	Delhi Surgical Dressings	4000–5000
4.	Silicone Taiwan Make	Rustagi Surgicals, Phoenix Medical Systems	2800–3500
5.	Silicone Korean Make	Delhi Hospital Supply	2400
6.	Silicone Ambu Make	Indian Surgicals Equipment	4000–5000

Size

Masks come in several sizes. Resuscitation tray should contain masks suitable for small premature infants as well as for full-term infants. For the masks to be correct size, the rim must cover the tip of the chin, the mouth, and the nose, but not the eyes.

- Too large a mask will lead to ineffective seal and possible eye damage.
- Too small a mask will not cover the mouth and nose and may occlude the nose.

Decontamination

Washing and rinsing: Thorough decontamination of the resuscitator is necessary, ideally after each use. Disassemble all parts. Wash thoroughly in warm water using a detergent that is compatible with the resuscitator materials. Rinse all the parts thoroughly in clean water. Dry them before reassembling.

Disinfection/sterilization: Disinfection is the process by which all the live organisms get killed while in sterilization even the spores are killed. Chemical disinfection can be done by soaking in 2% glutaraldehyde active solution for 20 minutes. Sterilization procedure takes at least 6 hours. One can use ethylene oxide for gas sterilization, for all disassembled parts of resuscitation, while boiling and autoclaving can be used for all disassembled parts of resuscitator except the reservoir. All parts should be dried before reassembling. If detergent disinfectant residuals are allowed to dry on the resuscitator parts, the surface may become sticky. This may cause valve malfunction. Carefully inspect all parts for damage or excessive wear and replace them, if required. Reassemble and test the bag for proper functioning. Common brands available in India are shown in Table 28.1.

Frequently Asked Questions (FAQs)

Q. 1. What should I look for before buying a resuscitation bag?
Ensure that it is capable of providing 100% oxygen, has a safety device inbuilt and it conforms to standard specifications (like patient outlet fixes to connectors of standard endotracheal tubes and other brands of masks).

Q. 2. What is the ideal capacity of a resuscitation bag for newborn?
Ideal capacity of a bag for a neonate is 240–750 ml. For a baby <1500 g use a bag of 240–350 ml capacity.

Q. 3. Can I use resuscitation bag for providing free flow of oxygen, if oxygen is connected to oxygen inlet?
Not all types of bags can be used for providing free flow of oxygen. Only bags with closed end reservoir or anesthesia bag may be used for this purpose.

Q. 4. How does self-inflating bag score over anesthesia bag?

Self-inflating bag is ready to use in emergency even if there is no supply of oxygen or pressurized gas. The concentration of oxygen can be varied with or without reservoir from 90–95% to 45–60% respectively. On the other hand, anesthesia bag always requires pressurized source of air or oxygen. If connected to oxygen it can deliver only 100% oxygen.

Q. 5. What are the indications and contraindications of bag and mask ventilation at birth?

Indications
a. Baby is apneic or gasping after initial steps of resuscitation.
b. Heart rate <100/minute.
c. Central cyanosis not improving with free flow of oxygen.

Contraindications
a. Meconium stained amniotic fluid with baby depressed at birth.
b. Congenital diaphragmatic hernia.

Q. 6. How can I judge for myself the amount of pressure I am able to generate with my hand?

You may train your hands and finger by simple test. Attach a long intravenous tube to patient outlet with an endotracheal tube connector. Let the tube dip to 15 cm under water level. Your fingers will have to generate pressure more than 15 cm, so as to cause air bubbles to be generated under the water column. Now let tube end sink to 20 cm below water level, you will have to generate more than 20 cm of water pressure. Similarly, one can judge for pressures of 25, 30 and 35 cm of H_2O, etc.

One can attach a manometer to patient outlet and read directly on a dial (1 mm of Hg = 1.3 cm of H_2O). Or if bag has a facility of pop off safety valve, one will have to exceed pressure of pop-off limit say 30 cm of H_2O when the hissing sound appears.

Q. 7. In an open-ended reservoir why is the tube corrugated?

Open-ended reservoir is provided with corrugations for increasing the volume of reservoir and when oxygen gets consumed from reservoir, it is drawn inside the bag in a laminar fashion.

Q. 8. What are ideal specifications of resuscitation bag?

Many locally made resuscitation bags are available which do not conform to ideal specifications. A few of them are highly unsatisfactory because of the poor quality of rubber, lack of facility for attachment of reservoir, absence of any safety features and loss of re-expansion of bag with use. Masks do not fit well at patient outlet and tight seal is difficult to obtain. Look for the following before purchasing a resuscitation bag for a newborn:

a. Capacity of bag (ideal 240–750 ml)
b. Provision for attaching reservoir
c. Safety device is present
d. Patient outlet is of standard size; endotracheal tube connectors and standard masks fit well into it
e. Easy to clean and disinfect
f. Withstands repeated autoclaving and boiling.

Q. 9. What is the function of valve adapter connected between air inlet and the closed-ended reservoir?

The valve adapter has the following functions
a. It regulates the pressure generated inside the bag. Once the reservoir is filled with oxygen the valve at air inlet and inspiratory valve at air outlet opens, so that continuous flow of oxygen is achieved. It results in PEEP of 2–3 cm of H_2O.
b. In case reservoir is completely full of oxygen, excess oxygen leaks from valve adapter to atmosphere.
c. In situations when there is no oxygen in the reservoir, while bag re-inflates air is drawn

in through the openings on valve adapter, thus delivering at least room air for resuscitation.

Q. 10. How often should one disinfect/sterilize bag and mask equipment?

This depends on number of babies needing bag and mask ventilation ensure that if it is used for a baby born following frank chorioamnionitis, the equipment needs sterilization before being used on next baby. In a busy hospital catering for 2000 births per annum, it may be a good idea to sterilize bag and mask every 15 days. But disinfection must be followed on daily basis. The mask must be disinfected after each single use.

Q. 11. What is sustained lung inflation?

During first few breaths of manual ventilation after birth providing initial inflation of 5–15 seconds at inflating pressure of 20–25 cm of H_2O. It increases the functional residual capacity, aeration of lung and stabilizes cerebral oxygen delivery. It is still restricted to randomized trials.

1. Daylight 4. Thermometer
2. Prism 5. Knob
3. Eyepiece 6. Adjusting screw for scale

Clinical Refractometer 153

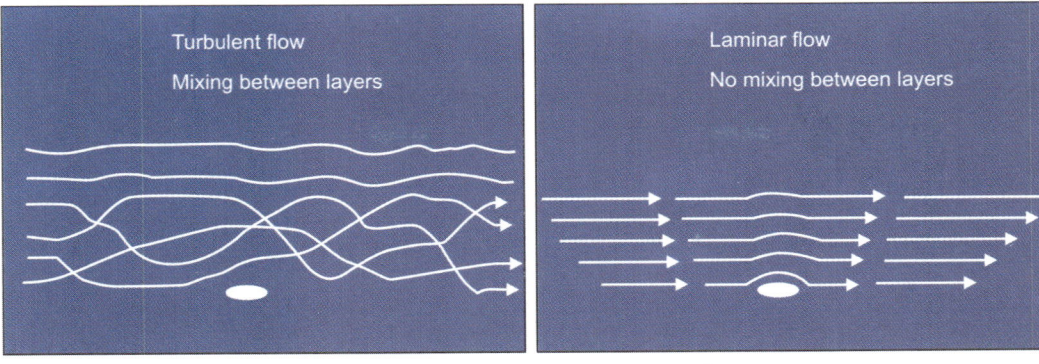

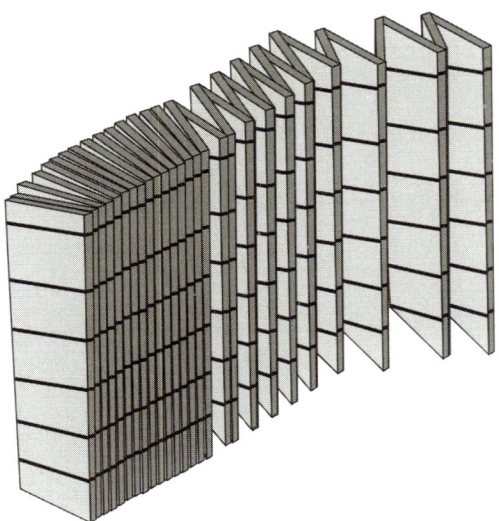

Fig. 27.2: Structure of HEPA filters showing the pleated structure of boron micro-fibers

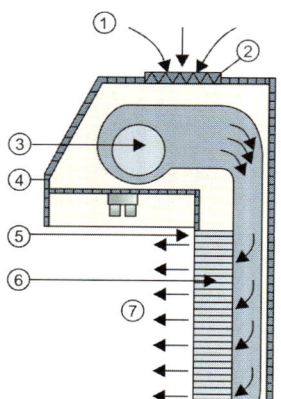

Chapter 29

Oxygen Concentrator

Oxygen therapy is life-saving in several respiratory and non-respiratory illnesses in both neonates and children. The most common indication of oxygen therapy in children is pneumonia, which is one of the leading causes of death in children less than five years old in most of the developing countries. Most of these deaths are associated with hypoxemia and oxygen therapy thus is an essential component of therapy. In developing countries, at district level, the source of oxygen is often the oxygen cylinder, and in some places liquid oxygen, but these are heavy, transport is cumbersome and need reliable distributing systems for refilling each time. The oxygen concentrator (OC) comes as a welcome change in place of these cylinders. Oxygen concentrators were first used in the 1960s to provide home oxygen therapy for patients with chronic lung disease.

A time tested device, mostly used in developed countries for domiciliary oxygen therapy, it has been actively promoted and field tested by WHO in developing countries as a source of readily available oxygen. They are 25–50% more cost effective than cylinders in resource poor settings. Although the initial expenditure of an oxygen concentrator is higher, in the long-term it is cost effective and though the initial cost of oxygen cylinders is low, the cumulative cost of refilling and maintenance is higher depending on transport and service costs in any given area.

CLINICAL USES OF THE OXYGEN CONCENTRATOR

1. As a source of oxygen in small hospitals for respiratory infections requiring oxygen therapy. Four children/neonates can be treated simultaneously using flow splitters.
2. In preterm neonates with chronic lung disease who are oxygen dependent at discharge.
3. In older children with chronic obstructive pulmonary disease/emphysema.
4. In children with ARDS with extensive fibrosis who may continue to require oxygen therapy for prolonged periods.

Principle of oxygen concentrator: Air contains a mixture of oxygen and nitrogen. The oxygen concentrator is about the size of a small refrigerator in which the air is forced under pressure through molecular sieve beds filled with Zeolite which binds the nitrogen (rapid pressure swing adsorption technology (PSA)), and separates it from air as in gas chromatography, thus increasing the proportion of oxygen from about **21%** to about **90%** (Fig. 29.1). This is how it works:

1. The air first passes through the four filters (explained below) which remove bacteria and dust.
2. The air is then forced by a compressor into canisters.
3. The canisters contain molecular sieve beds filled with zeolite (*aluminum silicate*) which binds the nitrogen and separates it from oxygen, thus increasing the oxygen concentration to as high as 90%.

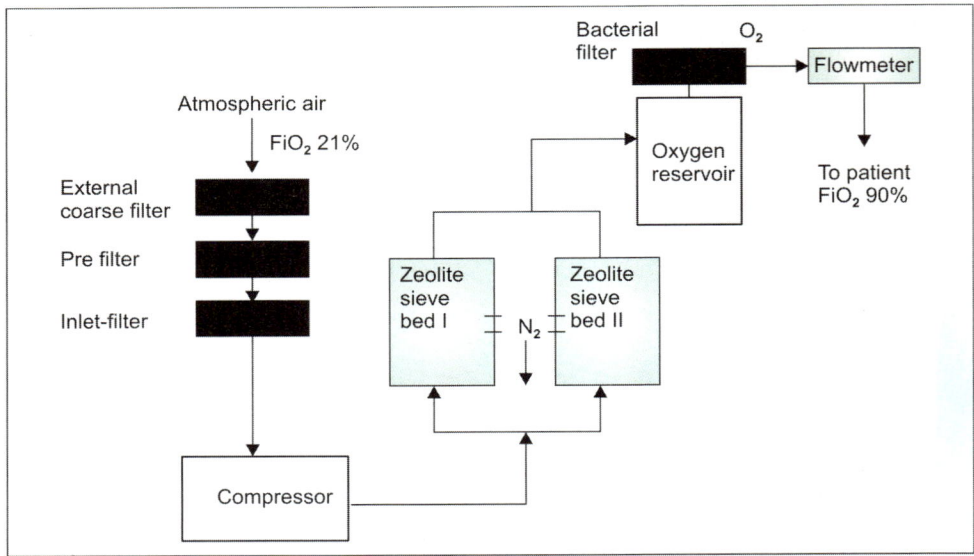

Fig. 29.1: Principle of working of the oxygen concentrator

4. Two canisters are used—the period of oxygen outflow from one coinciding with the discharge of nitrogen from the other, so that a continuous supply of oxygen-enriched gas is delivered to the storage vessel. As the first cylinder reaches near pure oxygen (there are small amounts of argon, CO_2, water vapor, radon and other minor atmospheric components) in the first half-cycle, a valve opens and the oxygen enriched gas flows to the pressure equalizing reservoir, which connects to the infant's oxygen hose. At the end of the first half of the cycle, there is another valve position change so that the air from the compressor is directed to the 2nd cylinder. Pressure in the first cylinder drops as the enriched oxygen moves into the reservoir, allowing the nitrogen to be desorbed back into gas. Part way through the second half of the cycle there is another valve position change to vent the gas in the first cylinder back into the ambient atmosphere, keeping the concentration of oxygen in the pressure equalizing reservoir from falling below about 90%. The pressure in the hose delivering oxygen from the equalizing reservoir is kept steady by a pressure reducing valve.

Components and accessories of the device

- There are three major controls—a power switch, an oxygen cock, and a valve for adjustment of flow. The power is supplied from mains, the consumption being 1 kW (240 V, 50 Hz).
- The concentrator (Fig. 29.2) has three filters: (i) External coarse filter; (ii) pre-filter; and (iii) inlet filter. Different concentrator companies have named the filters differently but standard names for the filters can be found in brackets beside the companies filter name.
- The air inlet and waste outlet are inside the cabinet and require adequate ventilation while the machine is running.
- Four children may be treated with oxygen at one time (at preset flows of 0.5 and/or 1 L per minute) using an oxygen flow splitting device.

Oxygen Concentrator

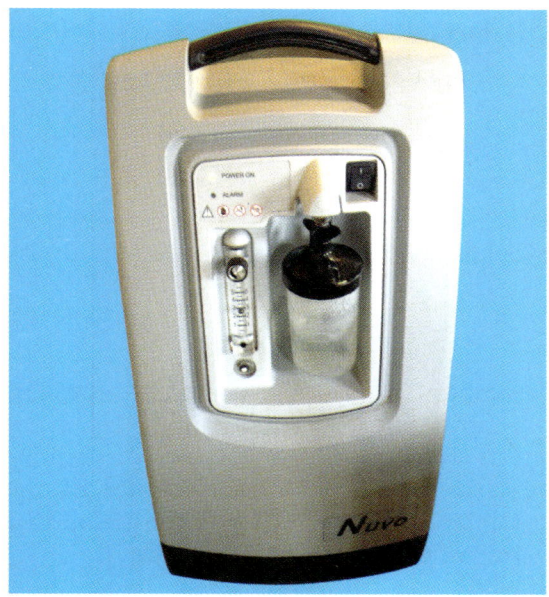

Fig. 29.2: Prototype of oxygen concentrator

- A "flow-splitter" (Fig. 29. 3) is a device connected to the outlet of an oxygen concentrator which provides four patients with oxygen at the same time. In order to provide oxygen to four infants it is necessary to set up four different nozzles (black arrows) for 1 L/min flow and yellow (white arrows) for 0.5 L/min flows. When less than 4 children are receiving oxygen then the unused outlets should be closed with blanking plugs. The oxygen is delivered through 10 to 15 meters of plastic tubing connecting the outlet nozzle and the nasal prongs fixed to the patient.

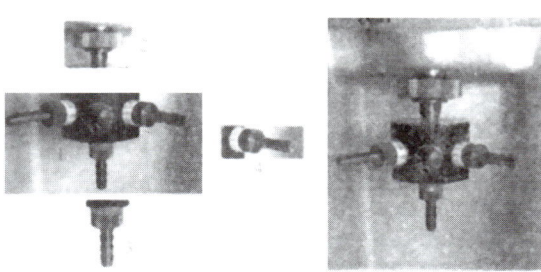

Fig. 29.3: Flow splitter

Device Specifications

The specifications of the wall mounted and the portable devices available in Indian market are described below.

Fixing the oxygen concentrator to the wall

The metal bracket is fixed with anchors and screws (Fig. 29.4, black arrow).

The oxygen concentrator is placed on the bracket and held in place by a moveable band of metal to secure it. This prevents the concentrator from being moved to patients instead of patients being moved.

Instructions of use of the oxygen concentrator

1. Plug in the power supply cable
2. Switch on the concentrator using ON/OFF button. Green power light will come on.
3. Adjust the flow rate to "4 litres per minute"
4. Adapt the flow-splitter and the calibrated nozzles or use blanking plugs as required.
5. The OSD (oxygen sensor device if present) should show a green light to indicate a normal concentration of oxygen (>90%).

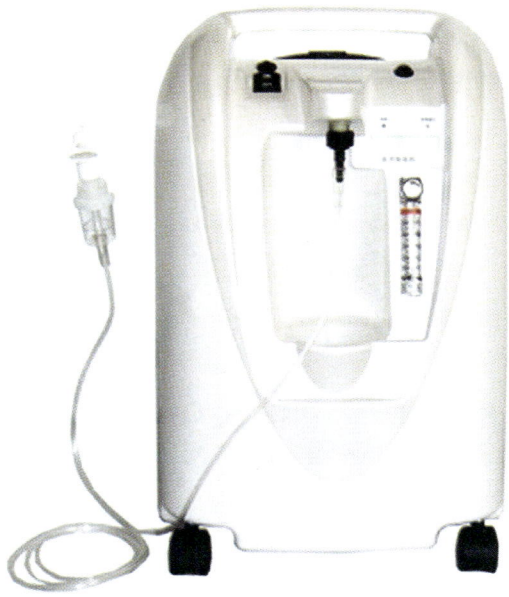

Fig. 29.4: Fixation of the oxygen concentrator to the wall

TABLE 29.1: Common oxygen concentrators available in the market

S.No.	Make	Dealer's name	Unit cost (₹)
1.	Devilbiss 525 SD	Devilbiss Healthcare	60,000–75,000
2.	Helix, Inspiron Tm	Helix Corporation, Bangalore	60,000
3.	Smart	Smart C International Ltd.	50,000
4.	Philips Respironics	Rustagi Surgicals, Phoenix Medical Systems	2800–3500
5.	Imported USA Invacare, Respironics, Airsep, Sequel	Respiroteh Med Solutions Pvt. Ltd.	1,00,000–1,50,000
6.	Newlife	Chart Industries, Inc.	40,000–50,000

6. Ensure that there are no air leaks
7. Make sure the nose is clear (saline nose drops)
8. Check the O_2 flow
9. Ensure that the nasal prongs are well-fitted to the patient
10. If pulse oximeter is available, monitor SpO_2 along with other vitals.

Common oxygen concentrators available in the Indian market are shown in Table 29.1.

Frequently Asked Questions (FAQs)

Q. 1. What is the maximum concentration of oxygen that can be achieved with the concentrator?
The maximum concentration of oxygen that can be achieved with the concentrator is up to 94%.

Q. 2. What kind of maintenance is required for the oxygen concentrator?
The device is as simple to manage as it looks. All that is required is that the coarse filter has to be washed each day and replace the other filters as per instructions of the company, which is usually at 6 months or 1500 hours of use. The filters may also need to be changed according to local conditions (dust, humidity) and the amount of use of the concentrator. Skilled mechanical maintenance is needed only yearly. There is a fourth filter, the bacterial filter, located before the oxygen outlet of the concentrator. Typically this must be changed every year, or more frequently, depending on the amount of use of the concentrator. The concentrator may work 24 hours a day, every day, as long as daily maintenance procedures are carried out. Every concentrator has an hour meter. The working hours should be written down in the service report.

Q. 3. How many infants can be treated at a time with the oxygen concentrator?
Up to 4 infants.

Q. 4. Are there any limitations of the oxygen concentrator?
In general oxygen concentrators have a few problems during use but it is important to be aware of a few limitations:
- When air is hot and humid, as may be the case in many tropical countries during summer season, the concentration of oxygen may be reduced to 70% because in these circumstances moisture is adsorbed by the molecular sieve material in preference to nitrogen.
- Low voltage may pose a similar problem by overheating the machine due to inefficient running of the motor. A voltage regulator should be used in these circumstances.
- At high altitudes (4000 meters), the oxygen concentration may be reduced to 80% due to low oxygen concentration in the air itself, but this should not cause serious difficulties in most cases.
- Small concentrators are not intended or suitable for compressed gas anesthesia (Boyle) machines or lung ventilators.

Q. 5. What are the precautions during usage of the oxygen concentrator?

The precautions to be observed are:
- The concentrator should be placed as far away from the window as possible to avoid dust and moisture.
- The concentrator should *not* be used to deliver oxygen through *head box* or *face mask*.
- It should be kept far away from open flames.
- Oxygen cylinders are recommended as a backup oxygen supply system in case of power failure as the oxygen reservoir in the machine lasts for only 2–3 minutes.
- When a flow-splitter is being used, the total flow should not exceed more than 4 L per minute or the concentration of oxygen will decrease. This happens because the canisters have a defined volume and can only separate a known volume of nitrogen at a given time. Therefore, concentrators must not be used at flows higher than those stated by the manufacturer.

Q. 6. How successfully has the oxygen concentrator been used previously?

Oxygen concentrators have been in use for more than two decades now for home oxygen therapy in patients with chronic obstructive airways disease, hypoxemia, and pulmonary hypertension in the adult age group. Considerably fewer children than adults receive home oxygen though these figures may be rising with improved neonatal care and increased number of successful preterm deliveries, thus increasing the overall burden of chronic lung disease in infants. Studies by the Medical Research Council and National Institutes of Health (USA) in such patients showed a substantial long-term reduction in mortality. To achieve this oxygen was required for at least 15 hours a day at flow rates sufficient to raise the arterial oxygen tension (PaO_2) to 60 mm Hg—usually around 2 LPM . Around four-fifths of patients using concentrators have chronic obstructive pulmonary disease and although there is no evidence of survival benefit among patients with other conditions, such as pulmonary fibrosis and bronchiectasis, long-term oxygen treatment by concentrator is often used if their arterial blood gas concentrations meet the criteria for this treatment.

Q. 7. Has the concentrator been tested as to what is the exact concentration of oxygen being delivered to the patient and how long this concentration is constant? Does the concentration of oxygen decrease over time?

The oxygen concentration measured both by an oxygen analyzer and by Scholander gas analysis was 92% when the flow was 2 L/min and fell to 81% at 3 L/min. After the build-up from 21% to 80–90%, which occurred over 20 minutes, the concentration of oxygen remained virtually constant for periods up to 16 hours, which was the longest time tested. From day to day the oxygen concentration at 2 LPM did not vary by more than 2%. The composition of the gas mixture delivered by the oxygen concentrator was first measured by the usual chemical methods (Scholander microanalyzer) and no carbon dioxide was detected. Further analysis by gas chromatography and mass spectrometry showed that apart from oxygen, the mixture contained nitrogen and argon, traces of carbon dioxide and water vapor but no undesirable component such as carbon monoxide or oxides of sulfur or nitrogen were detected.

Q. 8. What are the other methods of providing oxygen therapy?

The other methods of providing long-term domiciliary oxygen are cylinders delivered to the patient's home (the most widely used method) and liquid oxygen in a domestic tank replenished twice weekly in some countries. The capacity of the oxygen cylinder can range from 40 to 3445 L capacity and cost between ₹ 2,625 and 12,215. The cylinder can provide oxygen at 1 LPM for 11.3 hrs (e.g. 680 L capacity cylinder) or 4 LPM for 2.8 hrs. Thus, the overall running costs turn out to be much higher than the oxygen concentrator. The use of a liquid oxygen system has been investigated in the United States. The apparatus consists of a portable "walker" and a reservoir which contains a supply for three to four days. Regular deliveries are required as for cylinders and this contributes appreciably to the cost.

Q. 9. What is the running cost of the equipment?

The initial cost of the concentrator is ₹ 45,000–50,000, which may be considered expensive in comparison to cylinders. Nevertheless, the yearly running cost of ₹ 12,000–15,000 is substantially less than the ₹ 2.5 to 3 lakh for providing cylinders and the ₹ 1.2–1.3 lakhs for liquid oxygen. Thus, on cost alone the oxygen concentrator is the preferred means of providing oxygen treatment in developing countries with limited resources.

Q. 10. Does the oxygen concentrator require an oxygen sensing device?

Oxygen concentrator technology has improved rapidly and many concentrator companies are including oxygen sensing devices within the concentrator. The World Health Organization/UNICEF recommend that concentrators with sensing devices be purchased in countries where there is no other means of checking oxygen concentration. It is also prudent to have an oxygen analyzer for periodic checks (once every 3–6 months) of the oxygen concentration to be sure the sensing device is working properly.

Q. 11. What is the life of the zeolite sieve filters?

The life of the zeolite crystals can be expected to be at least 20,000 hours which in most situations would give about 10 years of use.

Continuous Positive Airway Pressure Machine

Continuous Positive Airway Pressure (CPAP) refers to the application of continuous pressure during both inspiration and expiration in a spontaneously breathing infant. By providing constant airway pressure, the alveoli are kept open and this increases the functional residual capacity (FRC) of the lungs resulting in better ventilation perfusion match and improved blood pH, oxygenation and carbon dioxide. It results in splinting upper airway thus decreasing airway resistance. These physiological effects result in decreased work of breathing and conservation of surfactant. Excessive amount of CPAP can result in pulmonary air leaks and increase in pulmonary venous pressure. This can adversely affect the brain (more risk of IVH) and heart.

An ideal CPAP delivery system consists of:
- A continuous supply of warm, humidified, blended gases at a flow rate of 2–3 times infants minute ventilation.
- A device to connect the CPAP circuit to infant's airway.
- Means of creating a positive pressure in the CPAP circuit.

CPAP DELIVERY SYSTEM (Fig. 30.1)

1. Ventilator

Ventilator is ideal system to provide CPAP but is expensive. It has blender for oxygen-air mixing, FiO_2 dial, humidifier, safety feature and a system to warm the gases. One simply has to switch over to CPAP mode and attach the infant.

2. CPAP System

CPAP system which delivers pressure and gases are available in India. An ideal system must have the following capabilities:
1. End expiratory pressure of 0–15 cm water
2. Humidification of up to 100%
3. Gas flow 5–8 L/min
4. Warming of gases to 34–37°C

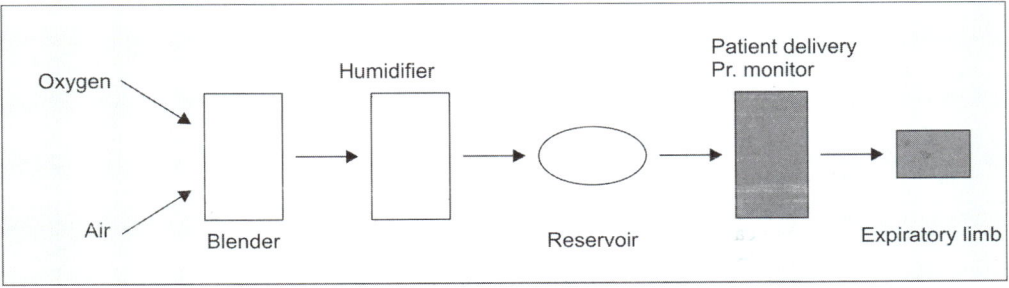

Fig. 30.1: CPAP delivery systems

5. Blending of oxygen air mixture (FiO_2 range of 0.21–1.0)
6. Display of FiO_2
7. Safety device against excess pressure
8. Tubing made of medical grade material with low dead space
9. Patient outlet to fit onto standard nasal prong system
10. Sterilizability of tubing
11. Low noise compressor
12. Capability to run continuously for days and weeks
13. Good esthetics
14. Easy maintenance
15. Reasonable cost

Unfortunately none of the Indian CPAP recently marketed in India have all the above features.

3. Improvised System

Several books show a simple CPAP system using under-water system. In principle this system is workable, albeit cumbersome. But, in India, where air—oxygen blenders are not available, it has to be used with only 100% oxygen. This means that we will always deliver 100% oxygen with CPAP which is not rational and can lead to development of retinopathy of prematurity (ROP) in sick preterm baby. It is, therefore, not a good method to be used in clinical practice.

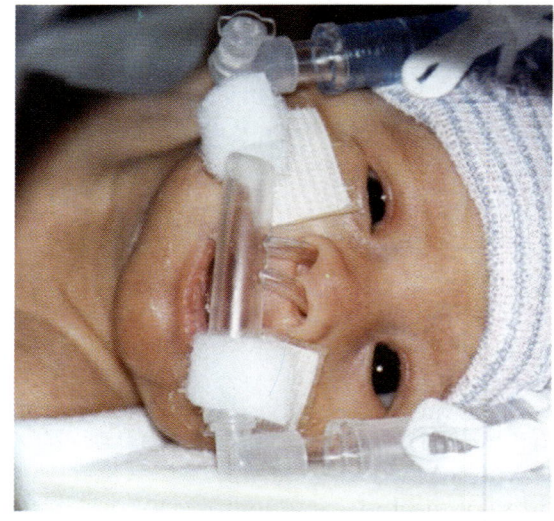

Fig. 30.2: Baby with nasal prongs with cap and circuit secured

Patient interfaces for providing CPAP (Fig. 30.2)
- Nasal prongs
- Nasopharyngeal prongs
- Nasopharyngeal tube
- Face masks

Nasal prongs are the most popular, convenient and practical method of providing CPAP. Nasopharyngeal prongs are also a convenient method but not so popular. Face mask CPAP is no longer in use.

Table 30.1 shows relative advantages and disadvantages of nasal prongs and naso-

TABLE 30.1: Advantages and disadvantages of CPAP devices

Feature	Nasal prongs	Nasopharyngeal tube
Advantages	• Simpler device • Ease of application • Mouth leak provides pressure relief	• Efficient • Fixation is easy • No leaks • Low gas flow can be used • High CPAP can be attained
Disadvantages	• Difficult to obtain good fit • Needs frequent oral suction • Difficult to apply in very small infant • Crying leads to loss of pressure • Needs high gas flow • May cause trauma to nasal septum/turbinate • May increase work of breathing • Liable to obstruction in prongs	• Invasive procedure • Risk of infection

pharyngeal tube as CPAP devices. The place of endotracheal CPAP (increases airway resistance significantly) is very limited in current neonatal practice.

Prerequisites for a good CPAP delivery system
- Flexible light weight tubing
- Ease of application and removal
- Low resistance
- Soft and relatively atraumatic
- Simple, easy to use and cost effective

Types of CPAP based on gas flow: CPAP is divided into two types depending upon type of flows, viz. Continuous flow and variable flow CPAP. Tables 30.2 and 30.3 show different machines and interfaces available in India.

Continuous flow CPAP: In this type a neonate is being provided a continuous fixed flow of gases irrespective of the phase of respiration at the expiratory limb of circuit. Ventilator derived CPAP, conventional standalone CPAP machines and bubble or water seal CPAP are perfect prototypes.

Bubble CPAP first described and used by Columbia University under the name Hudson CPAP in 70s. In this, warm and humidified blended gas flows to infant via nasal prongs. The distal end of the expiratory tubing is immersed in sterile water (addition of acetic acid makes the water bacteriostatic) to a specific depth to provide desired level of CPAP, e.g. the tubing is immersed to a depth of 5 cm to provide CPAP of 5 cm of H_2O (Fig. 30.3). Bubble CPAP with the combined effects of CPAP and pressure oscillations from the bubbles provides a lung protective, safe and effective method of respiratory support

TABLE 30.2: Available CPAP machines in Indian market

S.No.	Make	Dealers	Principals	Unit cost (₹)
1.	Indian CPAP	Phoenix Medical Meditrin Zeal Medical System Lectromedik Shreeyash	Phoenix Medical Meditrin Zeal Medical System Lectromedik Shreeyash	40,000 to 1,00,000/-
2.	Bubble CPAP	Fisher and Paykel Hudson Infant Pap Rohanika	Fisher and Paykel Hudson Galimed	1,60,000/-
3.	Fabian therapy Evolution with 2 in 1 (NIV and HFNC)	Rohanika	MS Acutronic Medical Systems AG, Switzerland	5,80,000/-
4.	Dual flow CPAP a. SIPAP b. IFD c. Arabella d. Medijet	 Rohanika Criticare India Pvt. Ltd. Trivitron, Chennai Pulmocare Consultants Phoenix Medical	 Sensor Medics, Viasys EME, UK Trivitron Medijet Phoenix Medical	1,20,000 to 3,00,000
5.	Fanhem	Fanhem, Brazil	Fanhem Medical Devices	1,60,000/-

Neonatal Equipment

TABLE 30.3: Patient interfaces available for CPAP

S.No.	Make	Dealers and address	Cost (₹)
1.	Argyle prongs	SB Medicare Pvt. Ltd. Mr JM Rishi Mb 9810117939 a. Telflex Medicals Pvt. Ltd. Blue Haven, N019	800/- available in three sizes <1 kg ; 1–1.5 kg; >1.5 kg
2.	Hudson prongs	Harrington Road, Chetpet, Chennai-6003031 919841722791/04428365040 b. Hudson RCI CNC Medical Devices, Mb 9820062869	1500/- six sizes with cap and patient circuit
3.	Nasopharyngeal prongs	Vygon India Pvt. Ltd. B 17/ Sector 34, Info City Gurgaon-122001 0124-4002801/802 www.vygon.com	500 for the prongs
4.	Fisher and Paykel	Fisher and Paykel 919849041888/ 918042844000	Nasal prongs ₹ 380/- Nasal tubing ₹ 800/- Caps ₹ 500/- Circuit ₹ 2200/-

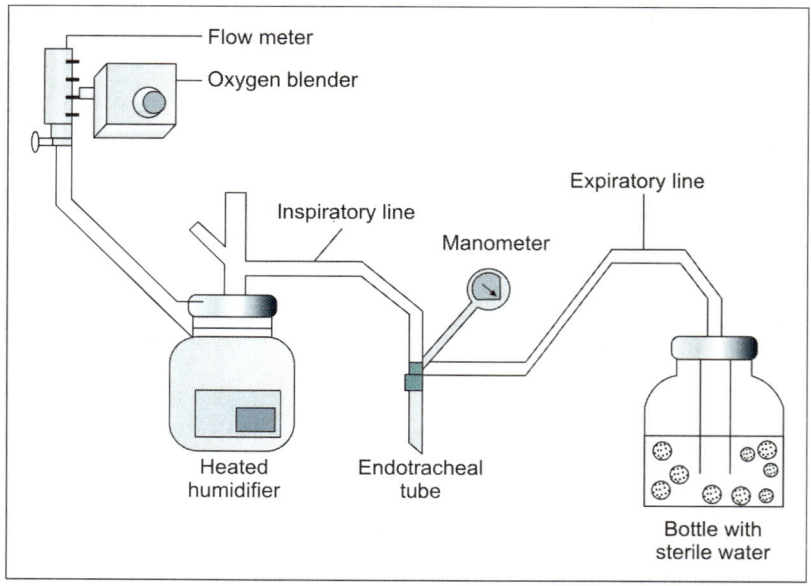

Fig. 30.3: Bubble CPAP

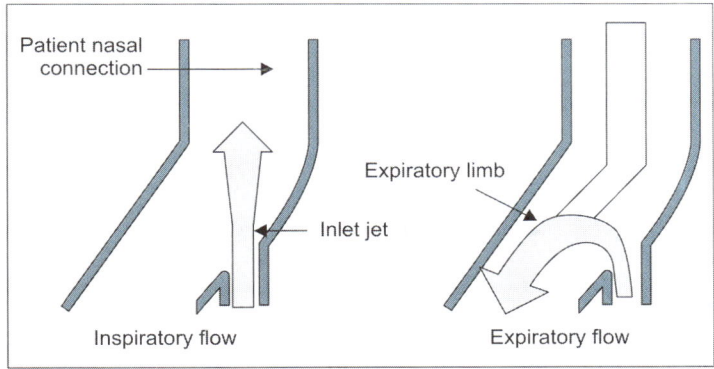

Fig. 30.4: Coanda effect

to spontaneously breathing neonates. The bubbling is also associated with the pressure oscillations, which get reverberated back into the infant's airway and may provide gas exchange through the principle of facilitated diffusion. This physiologic effect of bubble CPAP may help improve gas exchange and reduce the infant's work of breathing. Bubble CPAP may reduce the need for intubation, mechanical ventilation and reduce the incidence of chronic lung disease (CLD).

Hudson CPAP: Hudson prongs come in six different sizes for varied birth weight groups with cap and patient circuits with heated humidified wire. The prongs are applied using Tegaderm and Velcro; stabilized using safety pins on the woolen cap. The expiratory limb dips in the water bottle; the water level is the CPAP pressure in cm of water.

Variable/Dual Flow CPAP

Variable flow NCPAP generates CPAP at the airway proximal to the neonate's nares. It uses Bernoulli effect via dual injector jets directed towards each nasal prong in order to maintain constant pressure. If the neonate needs more inspiratory flow, the venturi action of the injector jets entrains additional flow. Due to Coanda effect during spontaneous expiratory effort there is fluidic flip which causes flow to flip around and to leave the generator chamber via the expiratory limb (Fig. 30.4). A residual gas pressure is provided by the constant gas flow, which enables stable gas delivery at a desired pressure during the entire respiratory cycle. The fluidic flip mechanism reduces the work of breathing almost to one-fourth of the continuous flow CPAP in which neonate has to exhale against the full continuous flow of gas. It is also found that variable flow maintains more uniform pressure level than continuous flow. The high cost and free availability are the major limiting factors.

Basic components of variable flow CPAP are
1. *Flow driver*: Provides adequate flows of appropriately blended gases
2. Humidifier
3. Flow generator
4. Patient's interface
5. Circuit
6. Fixation appliance

Examples
1. Infant flow system (Fig. 30.5)
2. Arabella
3. Sensor medics CPAP-SIPAP
4. Medijet (Fig. 30.6)

All four generators work on same principle of fluidic flip, have subtle differences.

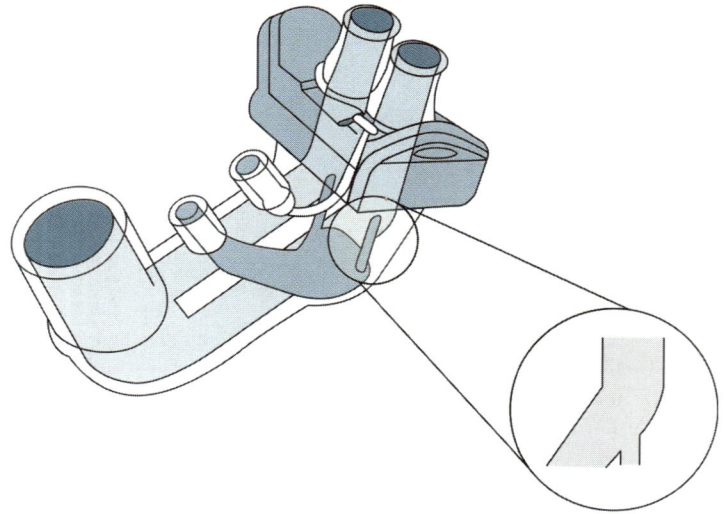

Fig. 30.5: Showing infant flow generator with fluidic flip effect

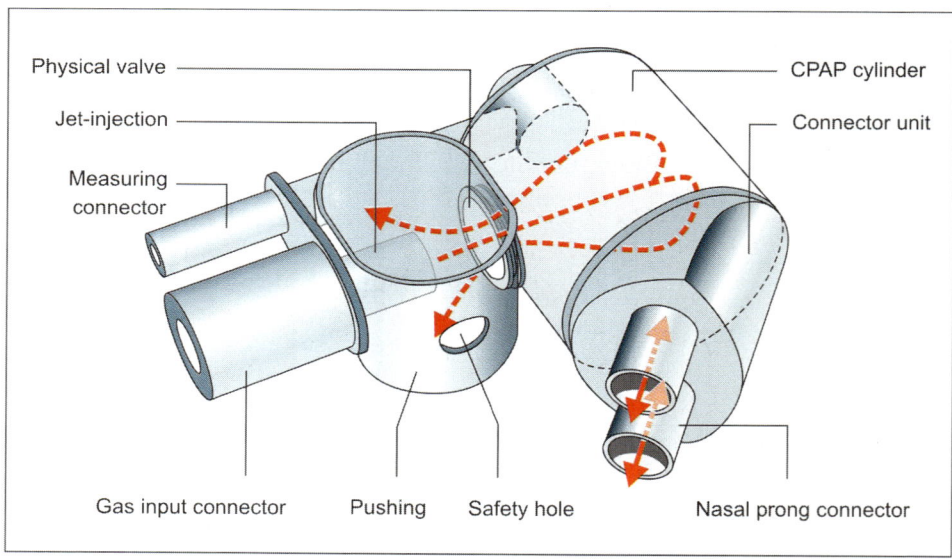

Fig. 30.6: Medical CPAP generator

Frequently Asked Questions (FAQs)

Q. 1. Can you convince me that one should use CPAP and avoid IMV/SIMV?
Intubation is the single major preventable factor contributing to bronchopulmonary dysplasia. It has been shown that even a single positive pressure breath can initiate the inflammatory cascade and cause lung damage. In addition, CPAP can conserve surfactant and maintain FRC. On the other hand, with intubation, we add lot of dead space and resistance—this leads to loss of FRC unless adequate PEEP is provided.

Q. 2. Is it essential to have a ventilator back-up in every unit with CPAP (to manage infants who fail CPAP)?

Ventilator back-up should be available but not necessarily in the same unit. If ventilator is not available in the unit, the unit should have arrangements for transport/pick-up and referral to another near-by unit with ventilation facilities.

Q. 3. How does indigenous CPAP fare when compared to either branded (F and P) bubble CPAP and ventilator derived CPAP?

Indigenous CPAP also needs a humidifier, and a Y connector to blend the air and oxygen. It will also need the interface. It should not be used with 100% oxygen or cold gases. Comparison of the two is not reported but as they work on the same principle, the ultimate pressure will be the same if correctly used.

Studies comparing the different systems have found that no single system is superior to others. Successful CPAP therapy is obtained by a proper fitting interface and having caregivers who are proficient in trouble shooting and taking care of a neonate on CPAP.

Q. 4. What are the practical problems/issues one would face initially while using indigenous CPAP?

Maintaining effective humidification and blending of oxygen and air are main bottlenecks.

Q. 5. I am planning to buy a CPAP machine. Should I purchase a machine providing CPAP alone or one with additional facility to provide ventilation? Should I go for an India-made CPAP machine?

A ventilator will cost at least 3 times that of CPAP alone machine. A dedicated CPAP machine will make the ventilator available for another infant needing IMV. So, it will be more cost effective to buy a stand-alone CPAP machine. India made CPAP machines are cheaper but have suffered from lack of generation of consistent pressures, reliable compressor and good quality interface. However, all the manufacturers have been constantly improving their models based on user feedback. Hence, it may be good idea to have a demo of the latest model being offered and test—use it before buying. Make sure that you are being provided a humidifier, reliable blender and a sturdy compressor (if you do not have central air supply) capable of working continuously for a week.

Q. 6. What are bare minimum needs for monitoring oxygen therapy on CPAP?

A pulse oximeter and a FiO_2 monitor. Keep the SpO_2 between 90 and 95%: For infants below 1 kg even saturations of 87% may be acceptable.

Q. 7. Is humidification a must for nasal CPAP as one is not bypassing upper airways?

Humidification and warming of inspiratory gases is the most important aspect of nasal CPAP. In normal breathing infants the flows are very low compared to the gas flows that occur when giving CPAP. High flow gases dry the mucosa, decrease mucociliary function and increase the airway resistance. It is ideal to deliver the gases at 37°C and at 100% relative humidity for best efficacy of CPAP.

Q. 8. Is it advisable to use CPAP without a blender?

No; one should not use CPAP with 100% oxygen. Either an electrical or mechanical blender or a Y connection to mix oxygen and air (in this case, FiO_2 monitor should be available) should be used.

Q. 9. CPAP is usually recommended for VLBW neonates with respiratory distress or post-extubation or for apnea of prematurity. Can I use CPAP for bigger or term neonates with respiratory distress?

CPAP is effective and can be used successfully in term infants with respiratory distress due to various etiologies like TTNB, MAS and pneumonia. Term infants may be more irritable and vigorous; hence keeping the prongs in place is more difficult.

Q. 10. How to select the appropriate size of the nasal cannula for infants with different birth weights?

Choose a prong that fits snugly for the nares. When using Fisher and Paykel prongs, follow the guide available with the kit. The guide consists of a transparent sheet with various sizes of nostril and columella printed on it as two horizontal rows respectively. Matching of the appropriate size of nostril with that of the columella can be done. The available possible combinations of these two measurements are indicated by lines joining the two rows. The best fitting size is then taken.

Argyll nasal prongs—sizes are related to birth weight of the infant. Three sizes are available one for <1000 gm, second for 1000–1500 gm and third one for >1500 gm.

Note that sometimes in ELBW infants, the prongs of a particular brand will not fit even though selected according to the weight/size criteria. In that case, try out the prongs of other brand and it will often fit (Table 30.4). Also, after one or two weeks, the nostrils often distend and one may have to change to the next higher size.

TABLE 30.4: While using the Hudson prongs

Weight of infant	Size of prong
<700 gm	0
700–1200 gm	1
1200–2000 gm	2
2000–3000 gm	3
3000–4000 gm	4
>4000 gm	5

Q. 11. What care should be taken of the prongs and nose while the infant is on nasal CPAP?

The biggest nasal prong, that comfortably fits the nostril, should be used. The weight of the prong should not fall on the nostrils. While selecting prongs it is important to select appropriate size so as to snuggly fit into the nasal cavity and have appropriate inter nares distance. While a tight one will cause pressure necrosis, the loose one will result in large air leaks. While inserting, it is important to see the symmetry and direction. The prongs should be moistened with saline before insertion. In a supine infant, the prongs should be inserted in the cephalad direction 15–20 degree downward to an imaginary perpendicular line drawn to body's long axis. The bridge of the prongs should not abut the columella and must not cause blanching of the skin.

The nasal prongs must be removed every 4–6 hours and nose inspected for secretions and normal saline should be instilled. The skin should be cleaned with sterile water and nasal prongs should be cleaned as per the manufacturers' instructions (with distilled water) every 4 to 6 hours.

It is very important to periodically inspect nasal mucosa for hyperemia; columella and alae nasi for blanching at least 4 hourly. If the prongs are required for >3–4 days it may be prudent to change them and sterilize by putting in glutaraldehyde solution before reapplication.

Q. 12. Can I use binasal flow oxygen cannula for providing CPAP?

It is difficult to obtain a tight fitting binasal cannula, so most often adequate CPAP will not be generated. These cannulae are meant for delivering low flow oxygen. However, if the outer diameter of nasal cannula is 3 mm and flows of 2 L/min are given, CPAP can be generated.

But note that nasal flow oxygen cannulae can generate very variable and unpredictable CPAP (even up to 8 cm of H_2O). Although CPAP can be generated, it has been shown that the work of breathing is increased because of high resistance. Hence, the ultimate chance of success will be less. In addition, adequate humidification will be very difficult to achieve with nasal cannulae.

On the other hand, short binasal prongs (Hudson, Argyle, INCA, IFD) being used as patient interface along with a pressure source have been demonstrated to provide best CPAP.

A device such as vapotherm had been used to deliver high flow nasal cannula CPAP but the experience is limited to a few centers in the world (refer to Chapter 31).

Q. 13. What is the status of providing CPAP via endotracheal tube?

Endotracheal CPAP increases resistance markedly, so the work of breathing will increase tremendously. In view of this, the Cochrane meta-analysis forbids use of ET CPAP as a weaning mode and recommends direct extubation from low rate IMV rather than after a trial of ET CPAP. Such a strategy increases the chances of successful extubation and also reduces chances of post-extubation apnea. In fact, in another review, Cochrane recommends extubation from IMV to nasal CPAP instead of head box oxygen as this reduces the chances of adverse events and extubation failure.

Q. 14. Binasal prongs are difficult to procure and are costly. Can I use an endotracheal tube inserted up to nasopharynx?

Nasopharyngeal tubes can provide effective CPAP and are easier to fix compared to short binasal prongs. However, in case of nasopharyngeal tube, the resistance is much higher and air leak through the contralateral nostril is likely to diminish the applied pressure. Moreover nursing care is impossible as there is no way to periodically monitor nasopharynx.

Better oxygenation, respiratory rates, and weaning success were reported with a short binasal device when compared with single prong nasopharyngeal CPAP.

Q. 15. How to select the appropriate size of the nasopharyngeal cannula, particularly when using cut ET tube—is it the same size as that for intubation or one size bigger?

Either the same size as that required for intubation or one size smaller.

Q. 16. How frequently should we do suctioning while using (a) nasal cannula and (b) nasopharyngeal cannula?

The suction is not routinely required in case of short binasal prongs. But periodic monitoring is must. If secretions are visible on periodic monitoring, gentle suction should be performed. In cases of nasopharyngeal prongs there is no way to inspect nasopharynx, so one has to perform suction every 4 hourly at least.

Q. 17. What is the best pressure to start with and how to regulate CPAP pressure?

For most lung diseases, the starting CPAP is 5 cm of water. For apnea of prematurity, the starting pressure is 4 cm.

When using bubble CPAP, increase or decrease in CPAP is done by increasing or decreasing the depth of immersion of expiratory limb in the bubble chamber. The water level should be kept constant.

Chest retractions, grunting and lung inflation <6 spaces on the X-ray are the indications for increasing CPAP pressure. Hyperinflated chest, lung inflation >8 spaces and flat diaphragm are indications for decreasing CPAP pressure.

After initiating CPAP, the decision to increase CPAP should be based on clinical signs. Presence of grunt, retractions and even tachypnea are indicative of inadequate support. SpO_2 values and blood gases can be useful adjuncts in decision-making in cases where mild tachypnea is present. One should not wait for X-ray to take decisions for increasing CPAP. FiO_2 should be adjusted to keep SpO_2 in range of 90–95%. CPAP pressure and FiO_2 generally go hand in hand, e.g. 5 cm H_2O:50%, 7 cm H_2O:70% and so on. If there is mismatch, e.g. FiO_2 30% and CPAP of 7 cm H_2O, this means we are wrong somewhere. X-rays are helpful to diagnose overinflation by the presence of >8 posterior intercostal spaces, flat diaphragm, tenting of ribs. During weaning as the FiO_2 requirements come down, CPAP should be brought down.

Q. 18. Besides the set pressure, what are the other factors which affect the delivered pressure to the infant on CPAP?

For a given set pressure, increasing the flow rate of the gases will cause an increase in the delivered pressure of the CPAP. So while changing the CPAP pressures on a given patient, the flow rate of the gases must be kept constant. The flow rate required depends on the CPAP generating system and other factors such as any leaks in the system and whether the mouth is closed or not.

The required flow rate can be objectively seen in bubble CPAP and rate can be adjusted to generate adequate underwater bubbles at the expiratory limb. For ventilator generated CPAP, the required exact flow rate is still debatable. A circuit flow of gases at 4–6 L/min may be sufficient if there are no leaks through the nose or mouth.

Q. 19. What parameters suggest that the infant is getting adequate CPAP while on bubble CPAP?

Comfortable infant, minimal retractions, good bubbling sounds in the chest, SpO_2 between 90% and 95%, continuous bubbling in the bubble chamber, minimal or no condensation in the inspiratory limb of the circuit and adequate chest expansion on the CXR.

Q. 20. Should the mouth be closed while giving CPAP?

Closing the mouth with a chin strap or pacifier will raise the pharyngeal pressure, whereas the pharyngeal pressure may fall if mouth is persistently open. However, open mouth also acts as a safety pop-off and helps to vent out any inadvertent increase in pressure delivered to infant. Hence, there is no need for routine closure of the mouth. However, if the infant is irritable and there is no bubbling in the bubble chamber even with flows >7 to 8 L/min, one may choose to close the mouth. It should be noted that most studies of CPAP have shown success without actively closing the mouth. It is important to realize that the neonates are obligatory nasal breathers. Most often, the cause of open mouth is pain or stress in the neonate. If these issues are addressed, majority of the cases will not need chin strapping.

Q. 21. Can I feed my infant on CPAP?

Definitely yes. We do not withhold feeds for an infant on CPAP. Most infants on CPAP in our units are managed successfully on full feeds.

The common fear is that the CPAP air will distend the stomach. But luckily, in premature infants the tone at upper end of esophagus and at gastroesophageal junction is higher than the usual CPAP pressures. So the air goes preferentially to the airways rather than the esophagus. Occasionally; 'CPAP belly', can occur; this is most of the times benign. Hence, while on CPAP, infants must always have an orogastric (OG) tube left *in situ* with other end open to atmosphere hanging with the canopy above the level of infant. This prevents efflux of feeds through the OG tube. Usually, the OG tube is plugged for 20–30 minutes after feeds.

Q. 22. What steps should I take to prevent common hazards and complications of nasal CPAP?

The major difficulty in using nasal CPAP is keeping the prongs in proper position. An interface that is light weight and is pivoted to the head and neck by proper fixation ensures that the nasal prongs do not get dislodged with the movement of the infant's head. Taking care of humidification of the gases, frequent suctioning and instillation of normal saline prevents nasal obstruction.

An inappropriately chosen large nasal cannula can cause nasal trauma including columella necrosis and long-term complication of nasal snubbing. A smaller size cannula may result in leak and CPAP failure. Hence, the correct size cannula needs to be chosen, following the guidelines mentioned for that product. Providing higher pressures than required can result in hyperinflation of the chest; if severe, it can result in air leaks (pneumothorax), retention of CO_2, reduction of systemic venous return and rarely impaired renal perfusion. Hence, continuous monitoring of infants on CPAP is a must to prevent these complications.

Heated Humidified High Flow Nasal Cannula Therapy

INTRODUCTION

'High flow' nasal cannula (HFNC) therapy refers to the administration of blended oxygen and air to newborn infants via nasal cannulae at higher flow rates than with low flow nasal cannulae (LFNC) (greater than 1 L/minute) (Cochrane 2011). Because flow used is high (2–8 L/min in infants and up to 40 L/min in adults), heated water humidification is necessary for avoiding dryness of nasal secretions and to maintain optimal nasociliary function.

In preterm infants, HFNC is used as a primary support for respiratory distress syndrome (RDS), for treating apnea of prematurity (AOP) and to support preterm infants after extubation and to wean from nasal continuous positive airway pressure (NCPAP).

MECHANISM OF ACTION

It is an open system in the sense that there is no expiratory limb. The gas which is delivered to the nasal surface is lost there itself. The proposed mechanism of action of heated humidifed high flow nasal cannulae (HHHFNC) therapy is:

1. It washes out nasopharyngeal dead space.
2. It provides adequate flow to support inspiration, there by reducing inspiratory work of breathing.
3. Improves lung and airway mechanics by eliminating the effects of drying/cooling.
4. It also believed to reduce metabolic cost of gas conditioning.
5. It can be used to provide continuous distending pressure, thus maintaining adequate functional residual capacity.

INDICATIONS

1. Respiratory distress as primary mode (rarely used)
2. Apnea of prematurity
3. For post-extubation support
4. Weaning from nasal CPAP.

CONTRAINDICATIONS

1. Congenital anomalies of nose like choanal atresia, etc.
2. Significant respiratory pathology.
3. Hemodynamically unstable patients.

EQUIPMENT AND PARTS

The basis parts of a HHHFNC include (Fig. 31.1)

1. *Humidifier*: For effectively warming and humidifying respiratory gases.
2. Respiratory circuit tubing (small volume circuit tubing) with a means to maintain the temperature and humidity, thereby preventing excessive precipitation or "rainout"—of the delivered gas until the distal end of the circuit.
3. A nasal cannula with adapter that connects it to the delivery circuit—size should be adequate to fit nares comfortably but not too tight to creat a seal (unlike CPAP where a seal is required).

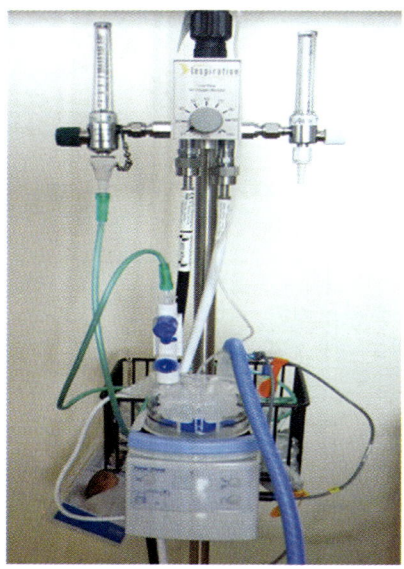

Fig. 31.1: Parts of heated humidified high flow nasal cannula system

Nsasal cannula size should not be more than 50% of the diameter of the nares.
4. Oxygen and air source, blender.
5. Flow meter—with flow range of 0–15 L.

PROTOCOL FOR INITIATING EQUIPMENT SET UP

1. First an appropriate size nasal cannula and circuit tubing is selected and nasal cannula is attached to circuit tubing. The circuit tubing is then connected to the humidifier. The temperature is set at 37°C in invasive mode so that temperature drop at nasal interface is in physiological range.
2. Attach air and oxygen compressed gas source to the blender and set adequate FiO_2 and the tubing from blender is attached to the humidifier. Set FiO_2 should be at same value, if infant is on another mode of non-invasive support or 5 to 10% higher if post-extubation. Maintain SpO_2 of baby at between 91 and 95%.
3. For initiating flow as a rule of thumb for initial flow rate setting use 1-2-3, 2-3-4, 3-4-5 formula, i.e.
   1 to 2 kg = 3 L/min,
   2 to 3 kg = 4 L/min,
   >3 kg = 5 L/min
   Increase flow rate in 1 L/min increments if
   - FiO_2 increases >10% above starting FiO_2
   - pCO_2 increases >10 mm Hg above baseline
   - Increased distress or retractions noted
   - Decreased lung expansion on CXR.
4. The inspiratory tubing containing humidified gas should be attached to appropriate size nasal cannula on patient side by using adhesive **wiggle pad** and make sure prongs do not tightly fit into nares.

PATIENT MANAGEMENT, FIO_2 MONITORING, HUMIDIFICATION AND NURSING CARE

Clinical Monitoring

A. Patient End

- Monitor patient for clinical response (respiratory rate, heart rate, degree of chest indrawing, and SpO_2).
- ABG and chest X-ray should be obtained as clinically indicated and not as a routine.
- Once stable on high flow, the infant is assessed regarding feed. They can be fed via orogastric tube end which should be kept open after 30 mins to an hour of feeding.
- Stable preterm infants on HHHFNC therapy can be put on breast to initiate breastfeeding and non-nutritive sucking.
- Kangaroo mother care (KMC) can be continued in babies while on HHHFNC therapy.

B. Interface

- Every 2–4 hourly nasal septum should be observed for erythema.
- Nasal suctioning should be done regularly as per babies need. Instill normal salene every 2–4 hourly nasal suctioning as per need.

C. Equipment End

- Humidifier water level as well as temperature should be checked regularly (at least 2 hourly).

Weaning

1. Clinically stable
2. Apnea free for last 3 days
3. Maintaining SpO_2 90–95% with FiO_2 <30%.

Side effects

1. Gastric distension
2. Nasal trauma (incidence less as compared to CPAP)
3. Air leak syndrome (Figs 31.2 and 31.3).

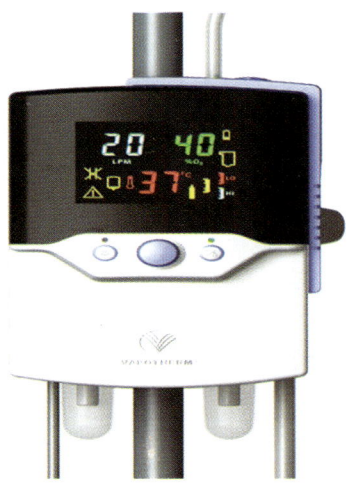 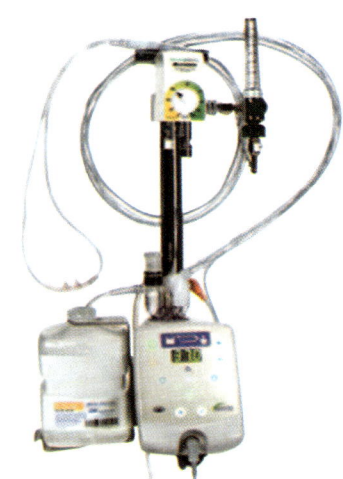

Figs 31.2 and 31.3: Prototypes of imported HHHFNC machines

Frequently Asked Questions (FAQs)

Q. 1. Does any Indian manufacturer make this in India?
No. All HHHFNC devices are imported from outside (Table 31.1).

Q. 2. Can a baby be put on KMC while on HHHFNC therapy?
Yes, KMC can be done while on HHHFNC. However, the neonate needs to be on continuous monitoring of oxygen saturation and heart rate during KMC.

Q. 3. What does evidence state regarding the safety and effectiveness of HHHFNC in preterm infants?
There is insufficient evidence to suggest about the safety or effectiveness of HHHFNC therapy as a form of respiratory support in preterm infants. The only benefit seen in studies was decreased nasal trauma. However, the absence of precision we still ensure of exact pressures delivery at nasal interface has limited its widespread use in India. We are still unsure of exact pressures delivered to baby.

Q. 4. Can HHHFNC therapy replace CPAP?
At present, evidence does not support use of HHHFNC as a primary mode and replacement to CPAP. It is bridge between CPAP and use of nasal cannula for oxygen in sick babies.

TABLE 31.1: HHHFNC machines available in Indian market

S.No.	Dealers	Principals	Unit cost (₹)*
1.	(AIRVO) Fisher and Paykel	Fisher and Paykel	2.5–3.0 lakh
2.	Vapotherm	Vapotherm	2.5–3.0 lakh
3.	Telefan, Chennai	Hudson	2.5–3.0 lakh

*Each disposable circuit will cost ₹ 2,500–3000/–

Chapter 32

Neonatal Ventilators

The advent of modern ventilator therapy in the 1980s has been a prominent milestone in the history of advanced neonatal care. In fact, it forms the sheet-anchor of any modern neonatal unit contributing significantly to lowering the neonatal mortality and morbidity of babies in respiratory failure. Ventilation should be undertaken in centers with adequate manpower, skill and technology. The nurse-to-baby ratio for a critically ventilated neonate should ideally 1:1 and definitely not less than 1:2. The choice of the ventilator, though a standard question, is not relevant. The conventional ventilators listed are all comparable with features like peak inspiratory pressure (PIP), positive end expiratory pressure (PEEP), inspiratory:expiratory (I:E) ratio, inspiratory time, mean airway pressure and can be optimally managed by any pediatrician skilled in ventilation. Lastly, but most importantly, ventilation should not be undertaken casually in centers where infection control is inadequate. It then becomes a self-defeating process.

BASIC PRINCIPLE OF A VENTILATOR

To understand the basic principle of a ventilator, we have to conceptualize the AYRES-T-piece principle. If you completely occlude one end, you produce inspiration by diverting gas flow to the baby. If the occlusion is released completely, gas flows through a path of least resistance and produces expiration. If the occlusion is released partially, pressure builds up at the end of expiration due to incomplete gas emptying. This is called positive end expiratory pressure (PEEP).

Mechanical ventilators incorporate a time cycling mechanism to occlude the expiratory limb of the patient circuit (inspiratory time), which may be pneumatic electronic such as Bear Cub, Sechrist or microprocessor controlled such as Infant Star.

The driving force of a ventilator is contributed by either compressed air or oxygen or both. A compressor giving 50–65 pounds per sq. inch (PSI) of pressure is to be used when a central source is not available. Oxygen is also to be given at 50 to 55 PSI. Once this driving force is obtained and the ventilator is operative, the concentration of oxygen and air can be adjusted by the blender to the required concentration (FiO_2). In addition, a flow meter regulates the flow between chambers A and B and prevents sudden transmission of pressure from (hypothetical) chambers A to B.

The inspiratory limb has a pop-off valve set by the manufacturer at 50–70 cm of H_2O as a safety measure for the baby. High pressure can be caused inadvertently due to sudden build up of pressure from chamber A to B due to high flow rates. A humidifier is also present in the inspiratory limb to humidify air to temperature close to the body temperature (37°C). Lastly, the inspiratory limb has a bacterial filter to sterilize the inspired air given to the baby (Fig. 32.1).

Neonatal Ventilators

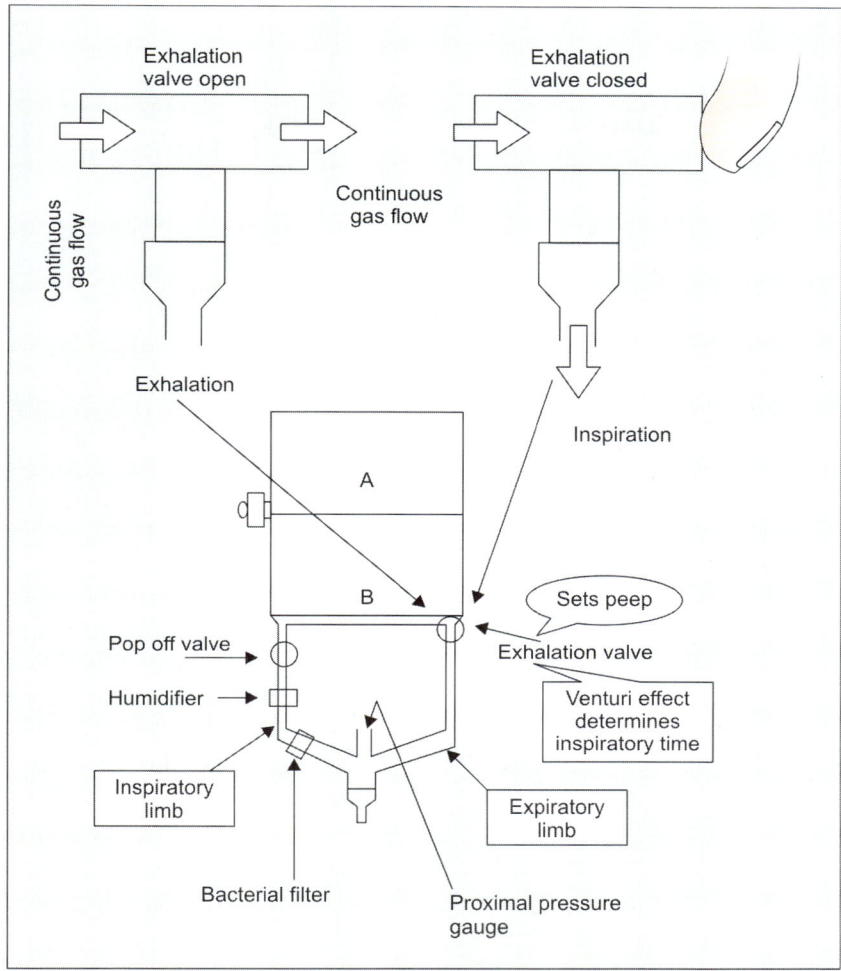

Fig. 32.1: Basic principle and components of a ventilator

The proximal pressure gauge measures the mean airway pressure (MAP), peak inspiratory pressure (PIP), positive end expiratory pressure (PEEP) at the distal end of the endotracheal tube and it does not truly reflect pressure in the alveoli at fast rates.

CLASSIFICATION OF VENTILATORS

1. By pressure relationship (negative or positive).
2. By cycling mode (time, pressure or volume).
3. By power source (gas or electrical).
4. By rate (high frequency ventilation and oscillation).

Ventilators can also be classified according to the control of ventilation (ventilation mode). To start inspiration, the machine may be triggered by the patient (assistor type), by the ventilator only (controlled type) or by both (assist-controlled type). In assist controlled ventilators, a device allows the patient to initiate some respirations but it also has a predetermined frequency of ventilator breaths (intermittent mandatory ventilation), as back up.

The current conventional ventilators in the market are time-cycled and pressure-limited, i.e. certain preset pressures (PIP) are reached at a preset inspiratory time (when the exhalation valve opens). In intermittent mandatory ventilation (IMV) machine's breaths is not always synchronized with patient's breath and it is possible that ventilator generated breath is superimposed on the expiratory phase of the neonate. Synchronized intermittent mandatory ventilation (SIMV) allows the patient's own breath so that superimposition of patient and ventilator breath do not occur. Ventilators incorporating high rates are called high frequency ventilators (HFV) and they often use tidal volume less than the anatomical dead space. Current HFV use rates between 150 to 2,400 bpm. They can be divided into HF jet ventilators (HFJV), HF oscillators (HFOV) and HF flow interrupter (HFFI) systems.

The basic requisites of a suitable ventilator for a neonate are (Figs 32.2 and 32.3):

1. It should be a *pressure limited, time-cycled ventilator and not a volume ventilator.*

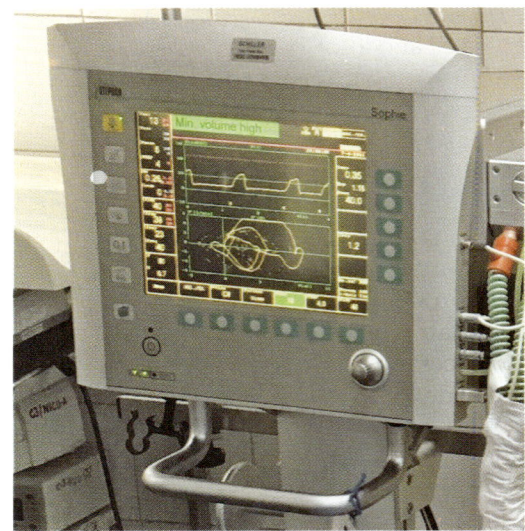

Fig. 32.3: Advance neonatal ventilator with graphics and HFO

2. All modes of ventilation should be available including *intermittent positive pressure ventilation (IPPV), continuous positive airway pressure (CPAP), SIMV* (if not too costly).
3. The ventilator should be simple and easy to operate.
4. It should have a minimum four-hour backup in an event of power failure.
5. Air-conditioning should not adversely affect the functioning.
6. It should be reliable and sturdy.
7. The machine should be relatively quiet, small and inexpensive.
8. The ventilator should offer a wide range of respiratory rates up to 150 bpm.
9. FiO_2 concentration should be adjustable and accurate (21–100%).
10. The ventilator should possess a low compilance system both inside and outside the ventilator.
11. The ventilator should provide accurate delivery of a wide range of tidal volumes (0.5–200 ml).
12. The ventilator should have a quick response time.

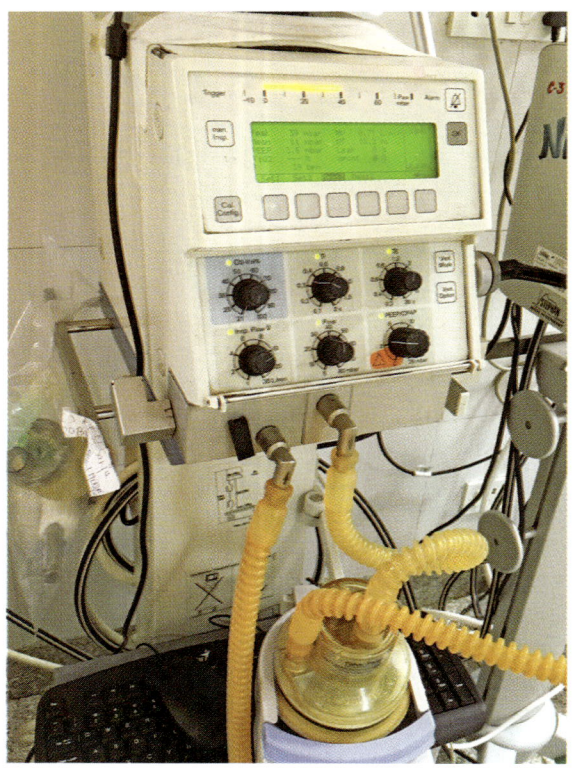

Fig. 32.2: Basic Draeger ventilator

13. The devices should have an alarm system (visual and audible) for mechanical failures, improper settings and blocked tubes.
14. The system should offer variable and constant flow rates.
15. The system should be capable of adequate humidification and heating of inspired gas (60% humidity at 37°C).
16. Optional: Ability to use nebulizer circuit for bronchodilator, etc.
17. Variable pressure or volume limiting devices should be available.
18. The devices should offer a wide range of pressure or volume capacities.
19. If possible, capabilities to upgrade the same ventilator to SIMV, HFV should be present.

ALARM SYSTEMS OF VENTILATORS

1. High inspiratory pressure (PIP > set PIP)
2. Low inspiratory pressure
3. Low PEEP/CPAP
4. Airway leak
5. Obstructed tube
6. *Insufficient expiratory time*: Activates when the ventilator rate and the inspiratory time settings are incompatible and results in an expiratory time shorter than the minimum allowed. The minimum inspiratory time for breath up to 100 bpm is 0.3 sec and for above 100 bpm is 0.2 sec.
7. Low O_2 pressure occurs when O_2 line pressure goes below the specified limits for that equipment.
8. Low air pressure occurs when compressor air pressure goes below the specified limit for that equipment.
9. Ventilator inoperative; occurs during system malfunctions or battery is fully depleted.

Maintenance Procedures

Protective Maintenance

A program of cleaning, operational preventive maintenance is important for the proper function and life of the ventilator. Using the quick checkout procedure on a routine basis will help identify potential problems.

Routine cleaning and checkout between patients should include:
1. Check out procedure as prescribed for the particular ventilator.
2. Water trap filter check.
3. Cleaning the exhalation block, diaphragm and patient circuit (if reusable), wiping the exterior of the ventilator.
4. Check air temperature of humidifier (37°C) and adjust control dials accordingly.

Monthly Operational Checks

1. Battery condition and LED display
2. Air and oxygen pressure gauge accuracy test
3. FiO_2 check
4. Proximal airway pressure gauge accuracy test
5. Driving pressure check

Yearly Preventive Maintenance

To be done two yearly or 10,000 hours of use.

Air and Oxygen

It is ideal to have a central supply of oxygen and air, each at with at least 50 PSI. There should be enough pressure to run all the ventilators in ICU; in addition, enough spare points should be available so as to allow manual ventilation with bag and mask. One can use mini manifold, in which cylinders are lined up in series. Large portable cylinders are cumbersome and they should be used in case above alternatives are not available.

Suction

Facilities of central suction supply would be ideal. Otherwise mechanical or electrical suction devices should be available at the bedside.

Power

Emergency back up is essential. Each patient's bed must have 8–12 electric points at a height of 5–6 ft (5 and 15 amp socket). The electricity supply should be on a stabilizer and it should be made sure that the load is not exceeded.

Cleaning and Sterilization

Follow instructor's manual for the equipment. Clean the exterior with mild detergent or a damp cloth. Do not let water enter the ventilator. Do not use strong solvents to clean the panels. The ventilator may be covered with cloth or plastic cover in case it is not in use.

It is ideal to use disposable patient circuits compatible with the ventilator (costing ₹ 2000/- to 2500/-). Reusable circuits should be immersed in glutaraldehyde solution for at least 6 hours, and then at rinsed in running water. Some tubing's which are autoclavable can be autoclaved. Gas sterilization for 4 hrs at 140°F with ethylene oxide and CO_2 or freon is ideal for a ventilator. Make sure the instrument is dried so that ethylene glycol is removed. One should have multiple circuits because they need sterilization every 2nd or 3rd day.

The filters of compressors should be cleaned in water daily. Bacterial filters should only be autoclaved. However, to extend its use, a culture of the bacterial filter can be taken after running on a test set up for half an hour. If the results are good, the filter is put to use. It is advisable to repeat the procedure monthly after 25 autoclaves.

Epilogue

Ventilator therapy is a serious undertaking where round-the-clock skilled care is needed. The choice of ventilator is not as important as the knowledge and experience in ventilation. Microprocessor based ventilators with battery back up and sensitive alarm system to indicate partially blocked tube are recommended. SIMV is definitely a useful ventilatory mode, especially during weaning a baby from the ventilator.

High frequency oscillatory ventilation (HFV) is a way of providing artificial ventilation of the lungs that theoretically may produce less injury to the lungs and therefore reduce the rate of chronic lung disease. However, HFV should not be used as a primary ventilatory modality in any center. IMV can be used to treat 90–95% of respiratory diseases. HFV is useful in pulmonary interstitial emphysema and air leaks and possibly in selected cases of persistent pulmonary hypertension. The bottom line is the skill and understanding of the ventilator and the mechanics of the lung disease.

Although many ventilators with some additional frills are available (making them more costly), it is advisable to start ventilation with a basic ventilator which is sturdy and has after sales services available. It might be quite relevant to stick to one type of ventilator for the case of maintenance and simplicity for nursing staff and junior doctors.

Frequently Asked Questions (FAQs)

Q. 1. What is synchronized intermittent mandatory ventilation (SIMV)?
SIMV is a mode where ventilator breaths are delivered at a fixed rate but are synchronized with the patient's own breath. However, unless the inspiratory times are identical the patient may start exhalation while ventilator is still in the inspiratory phase, resulting in partial asynchrony.

Q. 2. What is patient triggered ventilation mode (assist/control ventilation)?
PTV is a combination mode in which the ventilator delivers a positive pressure breath in response to the patient's effort (assist), provided it exceeds the preset threshold criteria. This mode also provides the safety of guaranteed mechanical breath rates as set by the operator (back up rate) if no patient effort is detected (control). The back up control rate ensures minimum mandatory minute ventilation in case the patient stops making an inspiratory effort.

Q. 3. What are the advantages of PTV?
PTV requires the least amount of patient effort and produces improved oxygenation at the same mean airway pressure. There is less

exposure to oxygen and lower pressures are required for ventilation.

Q. 4. What is pressure support ventilation (PSV) and how does it work?
PSV is defined as patient controlled ventilation which is generally flow-cycled. This is meant to assist the patient's spontaneous breathing with an inspiratory pressure 'boost'. In PSV, once the breath is triggered by patient's inspiratory effort, a preset system pressure is rapidly achieved and maintained throughout inspiration by adjustment of machine inspiratory flow. The inspiration ends when the inspiratory flow falls below a preset level.

Q. 5. Are there any advantages of using PSV mode?
PSV reduces the work of breathing created by the resistive forces of endotracheal tubes and the ventilatory circuit. It is mostly used as a weaning mode, but can be used as primary modality in patients with respiratory failure in presence of respiratory drive. PSV is better customized to support and synchronize with the patient's effort because the patient has control of both the inspiratory flow rate and inspiratory time.

Q. 6. What are the advantages of pulmonary graphics?
Pulmonary graphics are useful for improving quality of care for ventilated babies. Judicious use of graphics can optimize ventilation settings avoiding ventilator induced lung injury, pick up misadventures early like tube leakage, partial block, too much secretions and see the effect of therapy classical example is observing pressure volume loops after giving surfactant treatment to a premature baby with RDS. These add on comes at a cost but is often rewarding while managing ventilated baby.

Q. 7. Do we need high frequency ventilators in India?
Conventional ventilators are useful in nearly 90–95% neonates with respiratory failure. Rarely, in a situation of conventional ventilator failure, as in PPHN or when the baby develops pulmonary air leaks (pulmonary interstitial emphysema, pneumothorax) high frequency ventilator is indicated. But remember the cost of HFV is three times that of conventional ventilators. HFV would be justified in units which have more than 80% intact survival rates for ventilated babies between 1 and 1.5 kg.

Q. 8. Can I run a ventilator without a compressor, only on oxygen cylinders?
One needs to have compressed air at 45–55 PSI pressure from central source or a compressor to run a ventilator. Big oxygen cylinders, kept in series as 'mini manifold' outside NICU may provide oxygen source if centralized oxygen facilities are lacking in the unit. Remember, these cylinders will need frequent replacement depending on the consumption of oxygen. Compressor will still be required otherwise running ventilator on pure oxygen will deliver 100% oxygen.

Q. 9. What is the limitation of locally made compressors?
One must ensure that the compressor supplies medical grade air and uses oil free pistons. Most locally made compressors do not remove the humidity completely which damages the electronic pneumatic system of the ventilator.

Q. 10. Can one use modern ventilators safely in preterm neonates?
A few limitations need to be kept in mind: Firstly many fail to deliver small tiny tidal volumes and secondly pulmonary graphics may not be displayed appropriately. So always better to find out from others before investing on the costly ventilator.

Q. 11. What are the common brands of neonatal ventilators available in the Indian market?
The common brands of neonatal ventilators available in India are summarized below along with their costs and upgradability (Table 32.1).

TABLE 32.1: Transcutaneous monitors available in the Indian market

S.No.	Name	Dealers	Options special	High freq. upgradable	Cost per unit (in lakhs)
1.	Bear Cub BP-2001 Bear Cub 750 PSV	Instrument and Machine Life Care Medical Systems Rohanika	SIMV, PTV	—	₹ 5.5-8.0
2.	Bird VIP	Instromedix	SIMV, PTV	Yes*	₹ 8.0
3.	Babylog-8000 EVITA II	Draeger	SIMV, PTV	Yes*	₹ 7.0
4.	Event	Moola Tools	SIMV, PTV, Graphics	—	₹ 9.5
5.	SLE-2000 SLE-5000	Phoenix SBP Medicare	PTV, SIMV	Yes *	₹ 6.5
6.	Sechrist 100B Sechrist IV-200 SAVI	Pulmocare Consultant	IMV, SIMV PTV	— —	₹ 5.5 ₹ 8.0
7.	NMI E 100 Newport breeze	Pulmocare Consultant System Biomedical	SIMV, PTV	—	₹ 6.5
8.	Maquet Servo I/Siemens	Maquet Critical Care	PTV, SIMV, PSV, Graphics	—	₹ 10.0
9.	Sensormedics 3100 A, B	Life Care Medical Systems	—	Only HFO	₹ 13.0
10.	Stephanie Sophie	Ginevri	SIMV, AC, PSV, Graphics	Yes	₹ 14.0
11.	Aisys CS	GE Datex Ohmeda	PTV, PSV, Graphics	—	₹ 10.0
12.	Avea	Care Fusion	SIMV, PSV, AC, Graphics	Yes	₹ 15.0
13.	Fabian HFOV and Conventional	Rohanika	Equtronics, Switzerland		₹ 15.0
14.	HamiltonG-5	Hamilton, India	Hamilton AG, Germany		₹ 15.0

SIMV: Synchronized intermittent mandatory ventilation
PTV: Patient triggered ventilation
PSV: Pressure support ventilation
*Add ₹ 6 lakhs for this additional feature

Chapter 33

Inhaled Nitric Oxide Delivery Systems

Inhaled nitric oxide (iNO) therapy causes potent, selective, sustained pulmonary vasodilation and improves oxygenation in term newborns with severe hypoxemic respiratory failure and persistent pulmonary hypertension.

Multicenter randomized clinical studies have demonstrated that iNO therapy reduces the need for extracorporeal membrane oxygenation (ECMO) treatment in term neonates with hypoxemic respiratory failure.

The physiologic rationale for iNO therapy in the treatment of neonatal hypoxemic respiratory failure is based upon its ability to achieve potent and sustained pulmonary vasodilation without decreasing systemic vascular tone (Table 33.1). Distinct from its ability to decrease extrapulmonary right-to-left shunting by reducing PVR, low-dose iNO therapy can improve oxygenation by redirecting blood from poorly aerated or diseased lung regions to better aerated distal air spaces ("microselective effect").

BASIS (Table 33.1)

TABLE 33.1: Physiological basis of using inhaled nitric oxide in hypoxemic respiratory failure

1. Pulmonary vasodilatation decreased extrapulmonary right-to-left shunting
2. Enhanced matching of alveolar ventilation with perfusion (microselective effect)
3. ↓Inflammation (↓lung neutrophil accumulation)
4. ↓Vascular leak and lung edema
5. Preservation of surfactant function
6. ↓Oxidant injury (inhibition of lipid oxidation)
7. Preservation of vascular endothelial growth factor expression
8. Altered proinflammatory gene expression

INDICATIONS

1. In infants >34 weeks with hypoxemic respiratory failure with features of persistent pulmonary hypertension (PPHN) after adequate lung recruitment on high frequency oscillatory ventilation (HFOV), usually within first week of life, as evidenced by:

   a. Oxygenation index >25 on 2 ABGs at least 15 minutes apart.

   b. Echocardiographic evidence of PPHN, defined as right-to-left or bidirectional shunting at the ductus arteriosus or foramen ovale, or pulmonary artery pressure more than two-thirds of systemic pressure as estimated Doppler measurement of the tricuspid regurgitation jet.

   c. In the absence of echocardiographic facility, PPHN as evidenced by a difference of pre-ductal and post-ductal oxygen saturation >5% or more than two decreases in the arterial oxygen saturation to less than 85% in a 12-hour period despite optimal treatment of lung disease on HFOV.

2. In selected cases of congenital diaphragmatic hernia (CDH), with suprasystemic pulmonary vascular resistance after establishing optimal lung inflation and adequate left ventricular performance.

CONTRAINDICATIONS

1. Presence of major congenital anomalies
2. Grade 3/4 intraventricular hemorrhage

3. Echocardiographic evidence of a duct dependent cardiac condition such as hypoplastic left heart syndrome (HLHS), interrupted aortic arch, and critical aortic stenosis.

BASELINE STRATEGIES

1. iNO is only a part of overall clinical strategy in PPHN that cautiously manages parenchymal lung disease, cardiac performance and systemic hemodynamics.
2. Atelectasis, over distension and under inflation of the lung parenchyma should be avoided to optimize the effects of iNO.
3. Studies have shown that combination of HFOV with iNO produces maximum benefit than either alone in severe PPHN complicated by diffuse parenchymal disease.

INITIAL DOSING

1. The available evidence supports the use of doses of NO beginning at 20 ppm.
2. A brief exposure of higher doses (40–80 ppm) appears to be safe, but risks the development of methemoglobinemia.

RESPONSE

1. Babies showing adequate response (responders) show an improvement of PaO_2 >20%, fall in OI (OI less than 40 if initial OI >40) after 30 minutes of starting iNO.
2. In a study by Goldman et al, response to iNO was classified as nonresponders and responders.
   a. *The nonresponders* failed to show improvement in PaO_2 or OI even with 70 ppm of iNO for 10 minutes. The underlying causes were severe underlying lung disease with improper ventilation strategy or catecholamine refractory shock.
   b. *The responders* can be grouped into three categories:
      i. *Early failures*—failed to sustain this response over 36 hours, as defined by a rise in the OI to >40. 55% of these babies required extracorporeal membrane oxygenation (ECMO). Examples of this group are CDH, pulmonary hypoplasia secondary to oligohydramnios.
      ii. *Sustained responders* show initial and sustained response to both HFOV and iNO. All patients in this group survived and were discharged. The babies in this group were meconium aspiration syndrome, PPHN, and hyaline membrane disease.
      iii. Prolonged dependence—require initial high dose of iNO (40–50 ppm) and could be weaned to 10 ppm at 2–3 weeks and had sustained dependence on iNO for 3–6 weeks. All babies died in this group.

FAILURE

Clinically no improvement in saturations, color and perfusion status, or an arterial alveolar ratio a/A less than 0.22 defines failure of iNO + HFOV therapy, if iNO dosage reaches incrementally to a maximum of 80 ppm on high HFOV settings.

TOXICITIES

1. Methemoglobinemia resulting from reaction of NO with hemoglobin and has not been reported at lower doses of iNO (<20 ppm). If the levels are ≤5%, wean iNO by 50% and if levels are ≥10%, then it should be temporarily discontinued.
2. iNO combines with oxygen to form nitrogen dioxide (NO_2), a toxic gas. The safe upper limit of NO_2 is 3 ppm. If the levels are more than 3 ppm, the dose of iNO should be reduced by 50% and iNO should be temporarily stopped if NO_2 >5 ppm.
3. Theoretical risks of platelet dysfunction and bleeding problems. Risk of IVH reported in some studies done in preterm babies.

MANUFACTURE AND STORAGE OF NITRIC OXIDE

Nitric oxide is a small, diatomic, free radical that is unstable in the ambient atmosphere. It is highly soluble in lipid, reacts rapidly with oxygen (O_2) to form NO_2, which is potentially toxic. NO_2 is toxic at much lower levels than NO. The Occupational Safety and Health Administration (OSHA) has set exposure limits for NO in the workplace at 25 ppm time-weighted average for 8 hours. Cylinders for NO are constructed of an aluminum alloy. There are certain important characteristics of the storage space for cylinders.

1. Well ventilated to prevent accumulation of NO.
2. A means of securing cylinders and a means of securing the area to avoid unauthorized use.
3. Temperature conditions below 120°F.
4. Cylinders should be transported secured to a cart designed for moving cylinders and with a protective cap.
5. Full and empty cylinders should be segregated.

STEPS IN INITIATION (Fig. 33.1)

1. Evaluate the baby for the underlying causes of hypoxemic respiratory failure.
2. Ensure the baby has all monitoring devices in place including pre-ductal and post-ductal pulse oxymeters (preferably Masimo SET) and intra-arterial lines with invasive BP monitoring.
3. Document baseline heart rate, pre ductal and post-ductal saturations, blood pressures and ventilatory settings.
4. *Perform baseline blood tests*: Hemogram, platelet count and coagulation profile and methemoglobin levels (if possible).
5. Determine and document oxygenation index (OI = MAP × FiO_2 × 100/post-ductal PaO_2) in two ABGs at least 15 minutes apart.
6. Perform a baseline echocardiography (ECHO) to rule out duct dependent congenital cardiac lesion, evaluation of cardiac function, and diagnosis of PPHN.
7. Obtain baseline head ultrasound and neurological examination.
8. Optimize hemodynamic status, acid–base balance and sedation.
9. Re-evaluate the baby for the above mentioned indications and contraindications.
10. Connect the baby to the SLE 5000 ventilator with disposable tubings for iNO connection in the HFOV mode and optimize ventilation and ensure adequate alveolar recruitment.
11. Attach the NO_2 scavenger in the silencer portion of the ventilator.
12. Adjust the flowmeter in the front panel of the iNOSYS SLE system to deliver iNO at 20 ppm.
13. Assess the clinical response. Once connected to iNO system, do not change ventilator parameters unless urgently called for.
14. Obtain ABG ½–1 hour after initiation of iNO and then at least 6 hourly.
15. Consider weaning if criteria are met (see above). Obtain ABG ≤15 mins before weaning and then 30 mins after.
16. Consider escalation of dose or continued weaning to discontinuation, depending on the patient response.

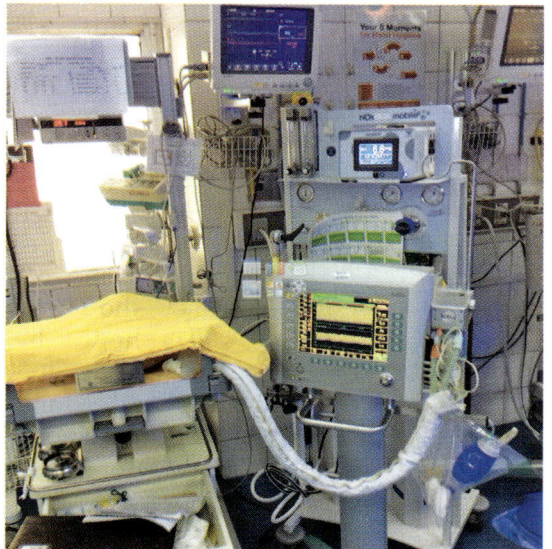

Fig. 33.1: Note baby on iNO therapy protected from sound and light

17. Monitor for rebounds after discontinuation.
18. Monitor NO₂ levels.

WEANING

1. Weaning of iNO is a different process than discontinuation.
2. During the process of weaning iNO, ventilator strategies and the FiO₂ are not altered.
3. Available strategies include:
   a. Reducing the concentration of iNO from 20 ppm to 6 ppm after 4 hours of treatment if neonate's condition is stable, PaO₂ at least 60 mm Hg with pH 7.4–7.55. If the above criteria is not met at 4 hours; iNO is continued at 20 ppm and evaluation is done every 4 hours till they are met and baby is reassessed.
   b. Reduction of iNO in a stepwise fashion by 50% of initial dose every 12 ± 2 hours. (e.g. 20 → 10 → 5 → 4 → 3 → 2 → 1 → 0 ppm). When this is successful, as defined as PaO₂ ≥125 mm Hg; further reductions are to be done till 0.5 to 1 ppm; from which, the babies will be discontinued.

ESCALATION OF INO DOSE

1. The maximum dose reported is 80 ppm. The risk of methemoglobinemia (blood levels >7%) and high inspired NO₂ (>3 ppm) are higher at these doses.
2. Besides, babies not responding to 20 ppm are unlikely to respond to higher doses. Hence, the maximum dose to be used is 20 ppm.
3. Escalation is done for babies deteriorating on the initial regime, defined as PaO₂<50 mm Hg or fall in PaO₂ >10% from the last wean.
4. It is done by doubling the initial dose of iNO (or to 5 ppm if the initial dose was <5 ppm) assessing response on ABG done after 30 minutes.

DURATION

1. The typical duration of iNO has been 5 days and this parallels with the clinical resolution of PPHN.
2. Longer duration may be required in certain cases such as pulmonary hypoplasia.
3. If iNO is required for >5 days, investigations into other causes of pulmonary artery hypertension must be considered, especially if discontinuation (see below) results in suprasystemic elevations of pulmonary arterial pressures by echocardiography.
4. The maximum cumulative iNO administration is 336 hours (14 days).

DISCONTINUATION

Plan to discontinue iNO if FiO₂<0.6 and PaO₂ >60 and if none of the following are present:
a. Evidence of rebound pulmonary hypertension
b. An increase of FiO₂ >15% after iNO withdrawal.

CONCLUSION

Inhaled NO improves oxygenation and decreases ECMO use in term newborns with PPHN. From the available information, a reasonable recommendation for starting dose of iNO in the term infant is 20 ppm, with reductions in dose over time. Toxicity is apparent at a dose of 80 ppm, which causes increases in methemoglobinemia and inspired NO₂. High doses (greater than 20 ppm) of iNO may prolong bleeding time, but clinically significant increases in bleeding complications have not been reported in term newborns. The use of iNO in non-ECMO centers must be done cautiously, with arrangements in place for transport to an ECMO center without interruption of iNO delivery in patients with suboptimal acute responses.

Finally, there is increasing evidence for the potential role of low-dose iNO (5 ppm) in premature newborns with hypoxemic respiratory failure. Low-dose iNO causes acute improvement in oxygenation and may prove to be useful as a lung-specific anti-inflammatory therapy; however, clinical application currently should be limited to controlled trials that target outcomes of both safety and efficacy.

SYSTEMS AVAILABLE IN INDIA

S.No.	Make	Dealers	Unit cost
1.	SLE INOSYS	GE Healthcare	10 Lakhs
2.	NOXBOXi	Rohanika	30 Lakhs
3.	NOXBOX mobile	Rohanika	30 Lakhs

TECHNICAL SPECIFICATIONS

1. Fully automatic and compact system to give controlled delivery of inhaled nitric oxide in the range of 1–99 ppm in increments of 1 ppm.
2. Should include dosing and analyzing unit, NO flow controller, calibration kit, NO gas cylinder with regulator and inline circuit sampling port.
3. The dosing and analyzing unit should be microprocessor controlled, use electrochemical monitoring with a response time <10 seconds; warm up time of less than 30 seconds.
4. The accuracy should be ± 1 ppm for NO and ± 0.1 ppm for NO_2. The dosing should be constant and independent of ventilator modes. It should use a flow proportional dosing.
5. Bright continuous and simultaneous display of NO concentration (0–100 ppm) and NO_2 concentration (0–20 ppm) in the inspired gas.
6. The audiovisual alarms should include: High NO_2; High NO; Low NO; Low gas supply; Tubing obstruction. It should have warning messages for calibration.
7. The unit should have a scavenging system to be attached to the expiratory limb of ventilator circuit.
8. Should have built in safety features such as pressure relief and safety valves fitted to both regulators.
9. Compact mobile stand and trolley to accommodate all components of the system.
10. The equipment should have compatibility with broad range of ventilators, including but not limited to Draeger Evita 2, Maquet Servo, Viasys Avea; the equipment should include all components necessary for delivery of NO using different ventilators; with manual bag resuscitator.
11. The equipment should work with standard and portable NO cylinders.
12. Calibration kit including gas cylinder, all connectors and tubing should be provided.
13. Standard nitric oxide (NO) cylinders 99.9% purity should be provided (cylinder should be filled).
14. The unit should run on 220/240 V, 50 Hz. Should also have an internal rechargeable long life battery, with a running time of at least 90 minutes.

Frequently Asked Questions (FAQs)

Q. 1. What are the indications of inhaled nitric oxide therapy?

Indications of use of iNO therapy are as follows:

a. In infants >34 weeks with hypoxemic respiratory failure with features of PPHN after adequate lung recruitment, usually within first week of life.
b. In selected cases of CDH, with suprasystemic pulmonary vascular resistance after establishing optimal lung inflation.

Q. 2. What is the contraindication of inhaled nitric oxide therapy?

a. Presence of major congenital anomalies
b. Grade 3/4 intraventricular hemorrhage
c. Evidence of a duct dependent cardiac condition such as HLHS, interrupted aortic arch, and critical aortic stenosis.

Q. 3. What is the initial dosing in inhaled nitric oxide therapy?

The available evidence supports the use of doses of NO beginning at 20 ppm.

A brief exposure of higher doses (40–80 ppm) appears to be safe, but risks the development of methemoglobinemia.

Q. 4. How would you define failure of iNO therapy?

An arterial alveolar ratio a/A less than 0.22 defines failure of iNO + HFOV therapy, despite giving highest permissible dose of iNO with high settings on high frequency ventilation.

Q. 5. What are the side effects of iNO therapy?

a. Methemoglobinemia resulting from reaction of NO with hemoglobin and has not been reported at lower doses of iNO (<20 ppm). If the levels are ≥5%, wean iNO by 50% and if levels are >10%, then iNO therapy should be temporarily discontinued.
b. iNO combines with oxygen to form nitrogen dioxide (NO_2), a toxic gas. The safe upper limit of NO_2 is 3 ppm. If the levels are more than 3 ppm, the dose of iNO should be reduced by 50% and iNO should be temporarily stopped if NO_2 >5 ppm. Limiting NO_2 production should be paramount in the design of any NO delivery system. Limiting residence time and using the lowest effective NO dose are 2 simple techniques toward achieving this goal.
c. Platelet dysfunction and bleeding problems may occur in iNO therapy. Risk of IVH has been reported in some studies done in preterm babies.

Q. 6. Is there any effect on the FiO_2 concentration of the inhaled gas after initiation of inhaled nitric oxide therapy?

From a practical standpoint, when delivering iNO through a mechanical ventilator, the clinician must decide how FiO_2 will be set and recorded. If FiO_2 is set to 0.6 and the addition of iNO reduces FiO_2 to 0.57, the clinician can record the measured FIO_2, record the set FiO_2, or increase set FiO_2 until measured FiO_2 reaches the ordered value. It is however believed that these changes in inspired oxygen are in significant with a good iNO delivery device and low NO dose, but these practical issues need to be addressed to assure uniform practice.

Q. 7. What are the available analyzers for detection of nitric oxide and nitrogen dioxide concentration?

Chemiluminescence analyzers determine gas concentrations by measuring stimulated photoemission. The NO chemiluminescence analyzer causes NO to react with ozone to produce NO_2 with an electron in an excited state. The advantages of chemiluminescence analyzers include high accuracy, ability to measure concentrations in the ppm range, faster response time, and greater specificity. Disadvantages include cost, size, the need for sample drying, and the requirement of a scrubber to eliminate ozone contamination from the work area. The electrochemical analyzers on the other hand are simple, have quick calibration and are cheap.

Section

H. Miscellaneous

34. Resuscitation Trolley
35. T-piece Resuscitator
36. Suction Machine
37. Breast Pump

Chapter 34

Resuscitation Trolley

Neonatal resuscitators with open care warmers are used in delivery areas for providing comprehensive care to a newly born infant. This includes temperature maintenance, suction, oxygen delivery and intermittent positive pressure ventilation when needed.

Basic Parts

In general, a resuscitator must meet the following basic criteria for maximum effectiveness and safety:

1. *Temperature maintenance*: The heater assembly consists of a radiant heater, parabolic reflector, observation light, and a visual and audible alarm system. The parabolic reflector focuses radiant energy on the bed surface minimizing energy loss due to scattering and provides an even field of radiant heat over the bed surface. The observation light provides intense light for procedures. Heater output of the unit can be controlled in two user-selectable modes—manual and servo-controlled. In the manual control mode, the heater output is set to a fixed level (in percentage of maximum possible output). In the servo mode of operation, user selects the patient's control temperature. A skin temperature probe is used to monitor the patient skin temperature. The control system modulates the radiant heat to maintain the patient at the selected control temperature. Infant's skin temperature, set temperature (in case of servo mode) and heater outputs are displayed on a light emitting diode (LED) panel. Audible and visual alarms activate to alert the operator of a low or high patient temperature, a skin temperature probe failure, a power failure, equipment failure or a check patient prompt.

2. *Oxygen delivery*: Oxygen is usually provided to the resuscitator from an oxygen cylinder or wall oxygen outlet. But the resuscitator must have provision for accurate delivery of oxygen at user selected flow rates. In addition, presence of a gas blender to regulate the concentration of delivered oxygen (FiO_2) is desirable. The accuracy of a gas mixture concentration is dependent on adequate oxygen and air supply to the blender unit. Therefore, an alarm must be provided if air and/or oxygen are not delivered at adequate pressure. Some resuscitators may have provision to use oxygen flow to generate suction (Venturi principle). This is especially helpful in case of a electricity failure.

3. *Suction unit*: Suction facility with inbuilt provision for monitoring the suction pressure is necessary. Suction pressure delivered by the machine can be generated either inside the machine (electricity driven or oxygen driven by Venturi principle) or provided through central wall mounted suction line. It is desirable to have an in-built back up facility of suction (oxygen driven) in the eventuality of electricity facility.

4. ***Positive pressure ventilation:*** Neonates who do not adequately establish breathing and/or have bradycardia require positive pressure ventilation. In addition to commonly used self-inflating bags (*refer* to Chapter 26), this positive pressure ventilation can also be provided by using T-piece. This device has a patient interface which attaches to a face mask or endotracheal tube. Inflation is achieved by interrupting the escape of gas through the outlet hole with a finger. A safety pop-off valve is usually attached. There must be provision for monitoring the delivered PIP (peak inflation pressure) using a manometer in the circuit. The length of the inflation is determined by the duration of the occlusion of the outlet. The *Neopuff Infant Resuscitator*™ (Fischer and Paykel, New Zealand) is a modified T-piece in which the outlet has a valve that controls the rate of escape of gas and can be adjusted to set a PEEP for a given flow of gas. The PIP is generated by occlusion of this valve with a finger (Fig. 34.1).

5. ***Timer:*** Presence of a count-up timer with audible signals at preset time intervals (for example 1, 5 and 10 minutes) is helpful in keeping track of the time elapsed during resuscitation. Timer function is used for decision-making during resuscitation and for assigning Apgar score.

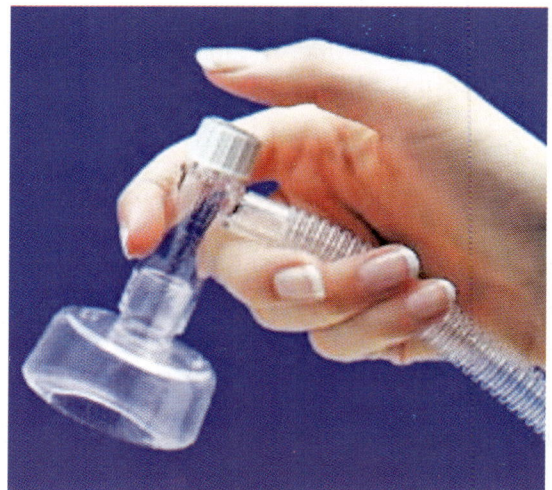

Fig. 34.1: Modified T-piece—Neopuff Infant Resuscitator™

6. ***Lighting:*** Illumination lamps are needed for adequate lighting during resuscitation. The lamps used must provide good color reproduction. This ensures detection of diagnostically important skin color nuances.
7. Mattress platform with removable side panels and facility for tilting the platform.
8. Rechargeable battery facility as a power back up is desirable.

Precautions

1. The apparatus must be inspected and serviced regularly by trained service personnel.

TABLE 34.1: Delivery room resuscitators available in Indian market

S.No.	Equipment	Company	Local dealers	Price
1.	CosyCot™ with Neopuff™	Fischer and Paykel, New Zealand	Fisher and Paykel India Ltd	₹ 2,50,000/-
2.	Atom Infa Warmer Model V-505™	Atom Medical Corporation	Vishal Surgicals	₹ 3,00,000/-
3.	Indian resuscitation warmer-free standing or wall mounted	Phoenix Medical System, Zeal, Meditrin, Shreeyash, Lectromedik	Phoenix Medical System, Zeal, Meditrin, Shreeyash, Lectromedik	₹ 30,000 to 1,00,000/- depending on available options
4.	Draeger Resuscitaire® Radiant Warmer	Draeger Medical, USA	Draeger Medical GE Healthcare	₹ 2,50,000/-

2. Follow user's manual for routine cleaning and disinfection of the instrument.
3. Do not leave the unit unattended when side panels are open.
4. Monitor the infant's temperature continuously on skin temperature display. Do not leave the baby unattended when heater output is in manual mode.
5. The oxygen concentration set by using the blender knob is only an approximation. For more accuracy, use an oxygen monitor to check.
6. Never use the T-piece resuscitator without a functioning pop-off valve and manometer.
7. Clean the suction jar periodically or whenever it is full of secretions.

Common brands available in India are shown in Table 34.1.

Frequently Asked Questions (FAQs)

Q. 1. Should T-piece be used routinely for providing positive pressure ventilation during resuscitation? What is the role of PEEP during resuscitation?

According to neonatal resuscitation guidelines (2005), a T-piece can be used to ventilate a newborn. A T-piece is a valved mechanical device designed to control flow and limit pressure. Target inflation pressures and long inspiratory times are more consistently achieved in *mechanical models* when T-piece devices are used rather than bags. Since the clinical implications of this are not clear, the *routine* use of T-piece for ventilation during resuscitation cannot be recommended at present.

Animal studies indicate that the inclusion of PEEP protects against lung injury and improves lung compliance and gas exchange. Application of PEEP may be beneficial when ventilating preterm infants after birth (Neonatal Resuscitation Guidelines, 2005).

Q. 2. Which warming mode to be used during resuscitation?

Manual mode should be used during resuscitation as the cot has to be pre-warmed before arrival of the baby.

Q. 3. Outline the routine for checking of resuscitation system by the staff.

A daily extensive check of all the functions must be made and recorded in a logbook. In addition, before each delivery, warmer, suction and oxygen supply must also be checked.

Q. 4. Enumerate steps to be used for use of resuscitator?

i. Before each delivery, the heater function of the resuscitator should be activated in the manual-mode. Heater output should be maintained at 60–80% of the maximum.
ii. Suction apparatus should be turned on. Suction pressure should not exceed 100 mm Hg.
iii. Oxygen supply should be opened to provide oxygen flow at a rate of 5–10 L per minute.
iv. Timer should be reset to zero reading.
v. If using T-piece for positive pressure ventilation, its functioning should be checked including in-built safety mechanism. During resuscitation, maximum pressure needed usually does not exceed 25–30 cm of water.

Q. 5. What routine maintenance is needed for delivery room resuscitator?

Manufacturer's recommendations on maintenance routines must be followed. Exterior of the resuscitator can be cleaned using mild detergent solution applied with a damp cloth or sponge. The mattress, X-ray tray, bed, and side panels may be cleaned without immersing by using a disinfecting agent safe for use on the materials (e.g. 2% glutaraldehyde). Clean the skin temperature probe by gently wiping with a soft, damp cloth containing detergent or disinfecting solution.

Chapter 35

T-piece Resuscitator

The T-piece resuscitator is a device that was first described (by Hoskyns, Milner, and Hopkin in 1987) as a tool to consistently provide prolonged inflations. T-piece resuscitator, a self-inflating bag and a flow-inflating bag are all acceptable devices to ventilate newborn infants either via a face mask or endotracheal tube.

A T-piece device requires a continuous gas supply to generate a set peak pressure and set positive end expiratory pressure (PEEP; Figs 35.1 to 35.3). T-pieces are easy to use, and are preferred by both experienced and inexperienced operators. In manikin studies, the T-piece device delivers peak and PEEP pressures that are more accurate and

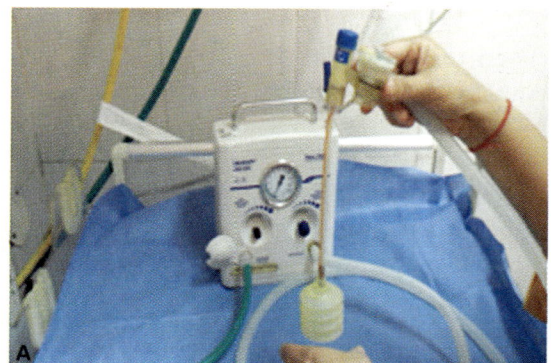

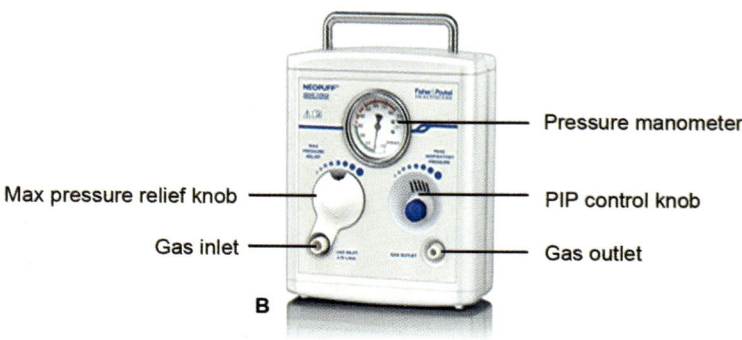

Fig. 35.1: T-piece resuscitator

T-piece Resuscitator

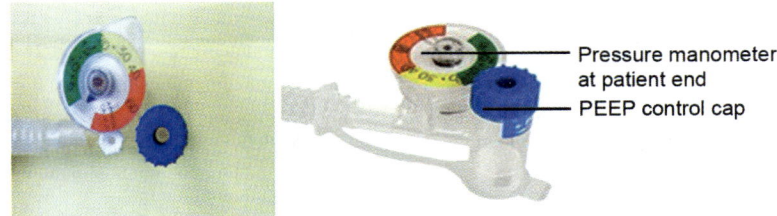

Fig. 35.2: Neo-Tee® (pressure manometer attached)

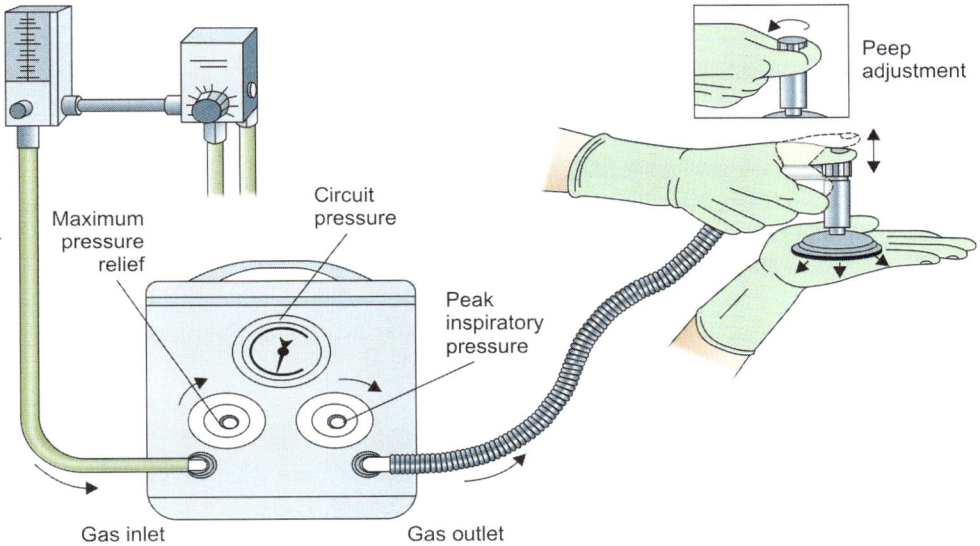

Fig. 35.3: Parts of T-piece resuscitator

consistent than other devices, resulting in more stable tidal volume delivery, even for inexperienced operators. However, care needs to be taken because alteration of gas flow during T-piece resuscitation changes the delivered pressures, particularly the PEEP. Occlusion of the hole on the T-piece allows delivery of a sustained inflation of any duration or pressure. The T-piece is capable of delivering CPAP and therefore may be the optimal device for providing respiratory support for preterm infants at birth.

SETTING UP THE T-PIECE RESUSCITATOR

To check and set the T-piece you will need:
- Infant T-piece resuscitator
- A compressed gas source
- Oxygen supply tubings
- Blender (preferable)
- A test lung (preferable)

Initial recommended settings are
- *Gas flow rate*: Set at 8–10 L/min.
- *Maximum pressure relief valve*: Set at 40 cm of H_2O.
- *Peak inspiratory pressure (PIP)*: Set at 15–20 cm of H_2O.
- *Positive end expiratory pressure (PEEP)*: Set at 3–5 cm of H_2O.

How to set
- Ensure pressure manometer read zero when no flow
- Set the gas flow rate to 8 L/min

- Connect the oxygen from compressed gas source to oxygen/gas inlet port on T-piece
- Connect T-piece tubing to gas outlet
- Attach a test lung to T-piece's patient end or can occlude with thumb
- *Set maximum pressure relief*: Turn the inspiratory pressure control dial fully clockwise until it cannot turn any further, occlude the PEEP cap on the patient T-piece, look at the manometer and open the cap covering the maximum pressure relief dial. Turn the maximum pressure relief dial clockwise or anticlockwise to adjust the maximum pressure to 40 cm H_2O (as maximum recommended for term babies is 40 and lower value about 30 for preterm), and then close the maximum pressure relief cap.
- *Set PIP*: Occlude the PEEP cap on the end of the patient T-piece.

  Turn the inspiratory pressure control anticlockwise several times to decrease the pressure from 40 cm H_2O down to recommended PIP.

  The set PIP is displayed on the manometer when the PEEP cap on the patient T-piece is occluded.
- *Set the positive end expiratory pressure (PEEP)*: Set the PEEP by turning the PEEP cap by clockwise or anticlockwise until the desired PEEP is displayed on the manometer.
- Remove the test lung and attach face mask or endotracheal tube to deliver set pressure by occluding the T piece cap hole at rate of 40–60/min.
- Occlude the PEEP cap using your thumb or finger for 0.5 seconds, then release for 0.5 seconds, it will provide rates of 60/sec.
- The peak pressure is achieved by occluding a hole in the top of the device with a finger. When the hole is not occluded, PEEP is delivered. Inflation time depends on the length of time the hole is occluded (but is often >0.5 seconds).
- While using T-piece and mask, good seal of mask is must to deliver adequate pressures. Leaks averaging 40–70% are common due to poor mask placement technique.
- *Therefore*
  i. Listen for a soft whistle of gas through the PEEP cap.
  ii. Look that a PEEP of 4–5 cm H_2O is displayed on the manometer.
  iii. Be aware that a PIP of 30 cm H_2O may be reached on the manometer **despite a face mask leak of up to 90%** (Wood *et al*, 2008).
- *Effective ventilation is confirmed by three signs*
  – An increase in the heart rate above 100/minute.
  – A slight rise of the chest and upper abdomen with each inflation.
  – An improvement in oxygenation.
- Some T-piece resuscitators are also available with in-built oxygen blenders (Lullaby Resus Plus, GE Healthcare).

Frequently Asked Questions (FAQs)

Q. 1. What are the advantages and disadvantages of T-piece resuscitator?
Ans. *Advantages*
  i. Provide consistent pressure
  ii. Reliable control of PIP and PEEP
  iii. Reliable delivery of 100% O_2
  iv. No operator fatigue from bagging

Disadvantages
  i. Requires compressed gas supply
  ii. Pressures to be set prior to use
  iii. Changing inflation pressures during resuscitation more difficult
  iv. Risk of prolonged inspiratory time
  v. 100% oxygen delivery may be harmful in preterm infants.

Q. 2. Why is PEEP so important?
Ans. Positive end expiratory pressure assists with lung expansion. It helps to establish functional residual capacity and improves oxygenation.

Q. 3. Does increasing the flow rate affects the delivered PEEP?
Ans. Yes, if flow rate has been increased, it increases the PEEP dangerously so need to re-adjust the PEEP.

Q. 4. What is the approximate cost of T-piece resuscitator?
Ans. Different manufactures have different costs, but approximate cost is 1.5 lakh.

Q. 5. What may be wrong if the baby does not improve or the desired peak pressure is not reached?
Ans. The possible underlying reasons may be:
a. The mask may not be properly sealed on the baby's face.
b. The gas supply may not be connected or flow may be insufficient.
c. The maximum circuit pressure, peak inspiratory pressure, or PEEP may be incorrectly set.

Q. 6. Can you give free-flow oxygen using a T-piece resuscitator?
Ans. Free-flow oxygen can be given reliably with a T-piece resuscitator if you occlude the PEEP cap and hold the mask loosely on the face.

Q. 7. Who are the various T-piece resuscitator dealers in India?
See Table 35.1.

TABLE 35.1: T-piece resuscitators available in Indian market

S.No.	Equipment	Company	Local dealers	Price
1.	Cosy Cot™ with Neopuff™	Fischer and Paykel, New Zealand	Fisher and Paykel India Ltd.	₹ 1,00,000/-
2.	Lullaby Resus Plus	GE Healthcare, UK	GE Healthcare, India	₹ 45000/-
3.	Baby puff 1020	Fanem, Brazil	Fanem Medical Pvt Ltd., India	—
4.	Resusibaby	Bird Meditech, Mumbai, India	Bird Meditech, India	₹ 28,000/-

Chapter 36

Suction Machine

Suctioning is used to remove secretions from the oral and nasopharyngeal area of the neonate using catheter to ensure patency. It is often used to prevent aspiration of oral or gastric secretion. The suction machine is useful in the labor room both for the obstetrician and the neonatologist. It is also essential in the neonatal ICU. A wall suction machine to suck out the secretions from mouth and nose are available in most hospitals. The maximum pressure should not exceed 100 mm Hg and this should be set before using the machine. It is important to ensure this especially if the same machine is being used for mother too. Any pressure higher than this is likely to damage the delicate mucosa and lead to iatrogenic problems. If suction machine is not available, a Delee's suction trap or a suction bulb may be used. Oral suction trap use is less advisable because of the risk of HIV and other infections from the secretions.

TYPES OF SUCTION SOURCES

1. Wall suction from a central source connected by pipes to various points in the hospital. This is expensive and possible only in big hospitals.
2. Portable suction machines.

a. Foot or Hand Operated

Foot Suction

Foot suction can be utilized to suck unwanted fluid from nasal and mouth suction. Various types of catheter can be attached for suctioning purposes. The foot/hand operated machines need physical effort, are less powerful but useful where electricity supply is erratic. When the bellows are compressed, air is forced through the air outlet valve in the upper part of the bellows. As the bellows re-expand, a vacuum is created and air is drawn through the tube connection out of the breaker. The vacuum in the bellows transfers the secretion from the patient to the beaker. The most effective result is obtained by a regular rhythmic compression of the bellows. Even if the beaker is full, the pump can compression will force the overflow to discharge through the overflow outlet in the bellows. Recent manual suction machines come with foot operated pedals which are compact, safe, effective and easy to operate. Vacuum pressure of up to 600 mm Hg can be created and the pressure generated can be monitored through the manometer. The jars made of polycarbonate are autoclavable and also are provided with bacterial filters beneath the lid or in the tube connecting the jar with the vacuum pump thereby ensuring sterility (Fig. 36.1).

Using the Foot Suction

1. Connect suction catheter to patient end of silicone tubing of machine.
2. Place the foot suction on floor across and in front of resuscitation trolley, with bellows on right side (if you use your right foot) and fluid collection jar on left side.

Suction Machine

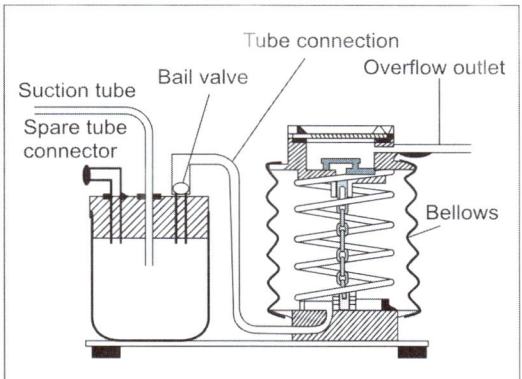

Fig. 36.1: Foot operated suction machine

3. Ensure that foot suction is close to resuscitation trolley so that it can be operated while resuscitating the baby.
4. Ensure that suction catheter is placed on baby mattress and tube length is not short.
5. Place right foot on bellows and press down ensuring that it slides down in contact with the central vertical metal plate. This ensures that bellows do not tilt outwards, preventing slipping of foot.
6. Foot pressure can be adjusted to ensure adequate suction pressure.
7. Pinching the suction catheter end press bellows and check for suction pressure.

   NB: For safety of newborn maximum suction pressure is limited to 100 mm Hg, irrespective of foot pressure.
8. In case suction inlet gets blocked by thick mucus plug, switch suction tubing to alternate suction inlet provided on rubber stopper.

Cleaning/Sterilization

1. The foot suction must be cleaned immediately after use. Empty the fluid collection jar.
2. The fluid collection jar and silicone tubing can be autoclaved at 124°C. This can also be washed with soap and water.
3. Wash the rubber stopper with soap and water and rinse thoroughly.
4. Re-assemble when dry
5. Replace in carry case.

   NB: Rubber lid for fluid collection jar cannot be autoclaved. Wash thoroughly with soap water, rinse and dry.
6. Empty fluid jar immediately when filled more than half.
7. In case fluid jar cannot be emptied immediately when full, to prevent overflow of fluid into bellow, open the alternate suction inlet. No suction pressure will be created even if bellow is compressed.

b. Electrically Operated

Components of the electrically operated machines:
1. Motor
2. Vacuum gauge with precision regulator
3. Jar(s)

Check before Purchase

1. There must be a tight seal of the vacuum jar. However, it should be easy to open the jar whenever it is to be emptied.
2. The jar should be easy to clean and sterilize.
3. There should be a regulator to adjust for the amount of suction required. During resuscitation at least 100 mm Hg suction pressure may be required. For slow suction which may be required continuously such as in a patient with tracheoesophageal fistula about 50 mm Hg may be required.
4. If lot of suctioned fluid or material is expected as during cesarean section or surgery, two jar suction machines may be required, but in most neonates where the amount of suction material is minimal, one jar is adequate.
5. Portable suction machine with oil immersed vacuum pump would produce less noise and would be appropriate in the NICU settings where unnecessary noise should be avoided.

Dual Suction Unit

Dual suction units are also available which can be operated manually as well as electrically and would serve the dual purpose of being useful with and without electricity (Fig. 36.2).

C. Venturi

Suction machines which are attached to imported resuscitation trolley run on Venturi principle. Air/oxygen flowing under pressure through a nozzle and opening in larger bore tubing in a sealed bottle create negative pressure inside the bottle. This machine is especially useful in case of electricity failure but must have continuous flow of oxygen to run on.

Common suction machines available in India are shown in Table 36.1.

Fig. 36.2: Typical dual power suction (electrical and manual)

TABLE 36.1: Common suction machines available in Indian market

S.No.	Make	Dealers	Cost (₹)
Mechanical			
1.	Foot/hand operated	Locally Babricated	600–800
2.	Ambu	Indian Surgical Equipment	5,000
3.	SU 701 suction machine, foot operated	Hospital Equipment Manufacturing Company	
Electrical			
1.	Local	Locally Fabricated	
2.	Atom D58	Vishal Surgical	40,000
3.	Atomos LC-16	Vishal Surgicals	30,000
4.	Medela (basic 1000 vaccum)	Rohit Hospital Supply	60,000
5.	Weyer Minivac ACU Model I	Rustagi Surgical	2,00,000
6.	Ameda	Medisphere	60,000
7.	Electric slow suction	Meditrin 20,000	
Dual (electric-cum-manual)			
1.	SU 05 suction unit dual power	Narang Medical Limited	80,000
2.	SU 503 dual power series (ASCO)	Apothecaries Sundries Manufacturing Company	80,000
Venturi			
1.	Fisher and Paykel	Fisher and Paykel Ind Ltd., Medisphere	2,00,000
2.	Atom V-3200 D	Vishal Surgicals	2,00,000

Frequently Asked Questions (FAQs)

Q. 1. What are the indications for suctioning of a non-intubated neonate?

- Presence of oral and/or nasal secretions with the infant unable to clear them on its own.
- Prior to bag and mask ventilation in meconium stained liquor.
- Presence of milk in airways.

Q. 2. What are the important points one need to keep in mind while undertaking suctioning?

- Avoid suctioning for 30 minutes to 1 hour after feeding unless it is necessary to establish patent airways.
- Endotracheal to be defered for at least 3 hours after instilling surfactant.
- Suction only when necessary. Routine suctioning increases risk of stimulation of vasovagal response which can lead to bradycardia and apnea.
- Do not exceed suction pressure of 100 mm of Hg (130 cm of water).
- Oxygen source and bag and mask should be available at bedside.
- Change the wall suction bottle and tubing everyday to minimize colonization with pathogenic organisms.

Q. 3. How much maximum suctioning pressure one can use in a neonate?

Not more than 100 mm of Hg which equals 130 cm of water.

Q. 4. Outline the procedure of suctioning.

See table below.

Q. 5. Is suction machine useful in any other circumstance other than suctioning?

In an infant with pneumothorax continuous low negative pressure of 10 to 20 cm water can be provided by connecting the suction tube to the one end of the under-water seal drain to which chest tube is connected. This allows rapid re-expansion of the lungs and to facilitate drainage of air. The same can be done in pneumopericardium and preoperatively for diaphragmatic hernic with a low pressure of 5 to 10 cm water.

Procedure	Rationale
1. Wash hands and wear gloves.	
2. Attach appropriate size catheter to suction tubing and insert catheter into sterile water	
3. Occlude catheter completely and set 100 mm of Hg (130 cm of water)	Prevents trauma to the mucosa pressure on suction caused by excessive pressure. Minimizes risk of hypoxemia and atelectasis
4. Estimate length of the catheter to be inserted by measuring from the tip of the nose to the tip of the ear lobe	Prevents catheter from reaching beyond oropharynx and stimulating the vasovagal reflex
5. Gently insert catheter to the measured distance from the mouth	Mouth suctioned prior to nares to decrease risk of gasping aspiration
6. Insert suction catheter gently upwards and back into the nares. If the catheter is difficult to pass, try with a smaller catheter. It is not necessary to pass a catheter completely through the nares to clear secretions (this may cause trauma) Applying suction to the external nares is often sufficient	This conforms to the direction of nares
7. Apply suction only upon withdrawal of catheter. Limit attempts to 3–5 seconds	Minimizes trauma to tissues and decreases risk of hypoxia or less
8. Rinse catheter in sterile water while applying suction and between suction attempts	Prevents occlusion of catheter
9. Gently insert catheter into one nares and apply suction	Forcing catheter leads to trauma
10. After suctioning, reposition the infant	
11. Discard catheter after single use	

Chapter 37

Breast Pump

Breast pumping is not meant to replace breastfeeding. It is to complement and enhance breastfeeding. Breast pump lets the mother collect her milk without stress and rush. Breast pumps are helpful in stimulating, maintaining and relieving milk supply in working mothers and mothers of premature or sick infants. Breast milk is produced on a supply and demand basis. Frequent pumping with complete emptying of breasts, encourages more production of milk.

INDICATIONS FOR USE OF BREAST PUMP

a. Feed a low birth weight (LBW) baby who cannot breastfeed.
b. Feed a sick baby who cannot suckle.
c. Relieve breast engorgement.
d. Feed a baby while he learns to suckle from inverted nipple.
e. Feed a baby who has difficulty in co-ordinating sucking and swallowing.
f. Keep up the supply of breast milk when the mother or baby is ill.
g. Leave breast milk for a baby when his mother goes out to work.
h. Prevent leaking when a mother is away from her baby.
i. Help a baby to attach to a full breast.

Mechanism of Action

The breastfeeding cycle is suck-release-relax (Fig. 37.1). Suckling action follows similar cycles of increasing suction, suction release

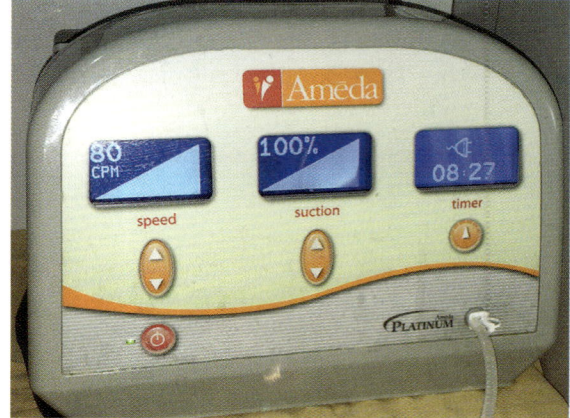

Fig. 37.1: A compact electronic breast pump

and relaxation when the suction is minimal and stable for a brief moment. Rapid nursing action of suck-release-relax cycle takes about one second. If babies feed nonstop, in a typical 20-minute breastfeeding session, they would complete about 1200 suck-release-relax cycles. Breast pump simulates a baby's suck-release-relax cycle to collect breast milk.

How does a breast pump create the physiological suck-release-relax cycle?

The three drawings indicate how a breast pump creates physiological cycle imitating a baby's nursing action.

1. *The relax phase*: The pump is at rest. Either the mother has just placed the pump on her breast to initiate pumping or she has just pulled the piston all the way in during a

normal pumping sequence. The breast shield and bottle are full of air and the breast is at rest (Fig. 37.2).

2. *The suck phase*: The piston is being pulled out. Because the system is closed, the air trapped inside the shield occupies a larger area. Therefore the air pressure in the shield decreases (suction is being created), causing the nipple to extend further into the shield. As a result of the decreased air pressure in the shield, milk is ejected from the nipple (Fig. 37.3). The ejected milk falls into the valve at the bottom of the shield inside the bottle. It is held there during the remainder of the suck phase.

Notice that the air pressure inside the bottle does not change during the suck phase. It is outside the closed system. It means that, as the bottle fills with milk, it has no effect on the amount of suction on the breast. In pumps where the collection chamber is part of the closed system, the amount of suction changes as the chambers fills with milk.

3. *The release phase*: The piston is all the way out. The rubber seal at the end of the piston has passed the vent in the cylinder. The system is no longer closed. Air is flowing back into the system through the vent and the pressure outside and inside the system is equal. Because the pressure inside the shield and the bottle are now the same, the weight of the milk collected in the valve can now push open the membrane and the milk falls into the bottle. The nipple returns to its shape (Fig. 37.4).

Stimulation of Milk Production

Breast pump is not only a collection device but it mimics the baby's sucking thus stimulating

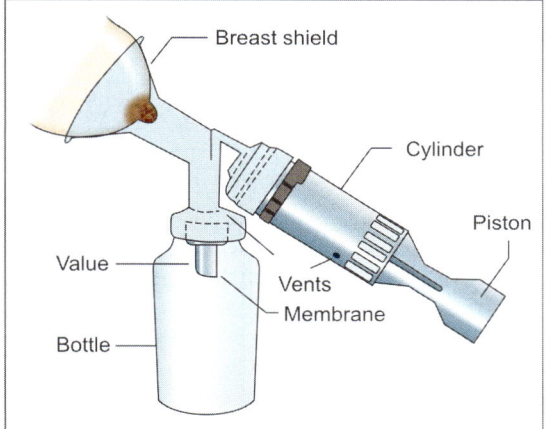

Fig. 37.2: Relax phase—piston all the way in. No suction. Breast and pump at rest

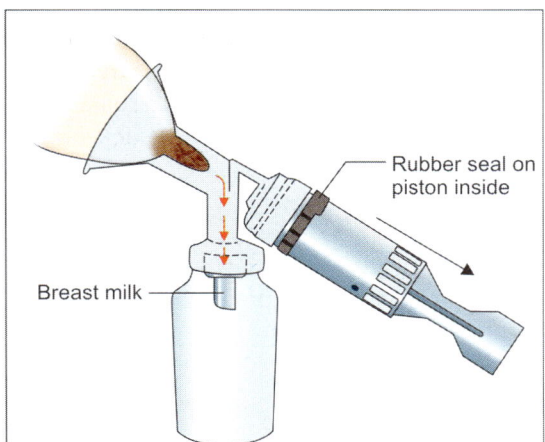

Fig. 37.3: Suck phase—piston being pulled out. Suction created on breast. Milk ejects

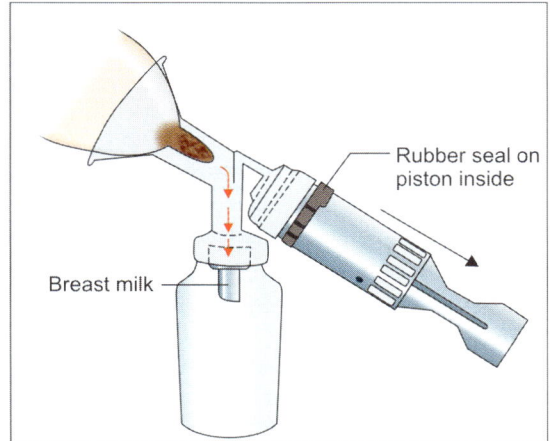

Fig. 37.4: Release phase—piston all the way out. Suction release breast returns to normal. Milk flows into bottle. Air flows back into shield when the rubber seal on piston passes vent

TABLE 37.1: Common brands of breast pumps available in Indian market

S.No.	Make	Principals	Dealers	Cost
1.	Medela a. Manual model b. Mini electric c. Lactina model	Medela Inc. P.O. Box 660, Mc Henry IL, 60051-0660	Phoenix Medical Systems	₹ 1800/- ₹ 8000/- ₹ 12,000/-
2.	Mee Mee breast pumps	www.meemee.in	Online	₹ 2000/- to 4000/-
3.	Avent comfort electric breast pump	www.philips.co.in	Online	₹ 12,000/-
4.	Lactec-10 St. Gallerstr. 23–25 Ch-9403 Goldach Switzerland	Nouvag AG Surgico Industries	International	₹ 22,000/-
5.	Mehar	Mehar Healthcare Corporation		₹ 800/-
6.	Medap	Medap	Rustagi Surgicals	₹ 48,000/-
7.	Minivac	M-Weyer Scheurener, Stra Be 90-5068, Odenthal	Rustagi Surgicals	₹ 50,000/-
8.	Ameda	Ameda	Medisphere	₹ 42,000/-
9.	BP-80	Atom	Vishal Surgicals	₹ 40,000/-

more breast milk production. Stimulation comes from the suck-release-relax cycles. A breast pump works physiologically and it automatically creates these suck-release-relax cycles. Typically it is created at approximately one cycle per second or second-and-one half. In manual pumps, the vacuum is released at the end of every pumping stroke thus completing suck-release-relax cycle. The rate of cycling is dependent upon how fast the mother is pumping.

Limits of Vacuum Pressure

Vacuum created on a normal pattern breastfeeding range from 0 mm of Hg during the rest phase of cycle to about 220 mm of Hg at the peak of suck phase, each cycle lasting for nearly one second. Majority of breast pumps work within safe, natural vacuum levels and durations. Unnaturally long or strong vacuum may cause pain or trauma to mother.

Hygiene

Collection device should be made up of such a material (polycarbonate), which can withstand boiling, or autoclaving. It should be kept clean before next use. Breast milk may get contaminated with bacterial growth. To prevent cross infection, parts of breast pump that come in contact with milk or mother should be sterilized before use. Breast pump should have a design so that milk does not enter electric motor in case of overflow or spillage.

Common brands in Indian market are shown in Table 37.1.

Frequently Asked Questions (FAQs)

Q. 1. How long can expressed breast milk be stored at room temperature?

It is safe to store freshly expressed breast milk up to 6–8 hours in room temperature (19–22°C). Afterwards, due to bacterial overgrowth, the milk is not suitable for feeding the baby. Even, if outdoor or room temperature exceeds 25°C, breast milk should immediately be chilled to preserve freshness.

Q. 2. Can breast milk be stored in refrigerator?

Expressed milk can be refrigerated (at 0°C) in sterile glass or plastic container for 24 hours without increase in bacterial contamination.

Freshly expressed breast milk can also be stored in vaccine carrier at 15°C with three frozen ice packs for 24 hours. In the freezer compartment of refrigerator, breast milk can be stored for approximately 3–6 months and in a separate – 20°C freezer, breast milk can be stored approximately 6–12 months.

Q. 3. How to warm/thaw frozen breast milk?

Place the sealed container of frozen or refrigerated breast milk in a bowl of warm water for 30 minutes. Once thawed, gently shake the container to blend the separated fat if any and do not store that milk again. Feed thawed milk immediately but test the temperature before feeding the baby. Do not thaw or warm breast milk in microwave oven.

Q. 4. How can one decrease bacterial contamination in expressed breast milk?

Careful handwashing is critical, and the nipples should be wiped with cotton and plain water before the milk is expressed. The first 5–10 ml of milk contains large number of bacteria; discarding this portion greatly decreases the contamination of expressed milk.

Q. 5. What standard microbiologic guidelines one can follow safely for expressed breast milk?

No standards of the microbiologic quality of expressed milk are available. Each millimeter of expressed human milk contains 10^3–10^4 colony forming units of normal skin bacteria, such as *S. albus* and diphtheroids; this milk can be fed without any ill effects. The presence of gram-negative rod in the milk indicates a problem in the collection technique. Milk containing more than 10^2 colony-forming units of gram-negative bacteria per ml lead to feed intolerance, and higher levels have been associated with sepsis.

Q. 6. Can heat treatment be resorted to for decreasing bacterial contamination?

Bacteria levels in milk can be controlled by heat treatment, which involves heating to 56°C for 30 minutes. This heat treatment leads to a 15% loss of secretary immunoglobulin A, a 25% loss of lactoferrin and folate, a 75% loss of phosphatase, and total elimination of beneficial cellular elements.

Q. 7. Does routine culture of expressed breast milk help?

Routine screening of EBM for bacterial counts is not warranted. But if there are concerns about expression technique and when intolerance by infant is suspected it may be done.

Q. 8. How does one express breast milk by pump?

Syringe pumps are more efficient than rubber bulb pumps. They are easier to clean and sterilize. Never use rubber bulb pumps.

1. Put the plunger inside the outer cylinder.
2. Make sure that the rubber seal is in good flexible condition.
3. Put the funnel over the nipple.
4. Make sure it touches the skin all around to make an airtight seal.
5. Pull the outer cylinder down. The nipple is sucked into the funnel.
6. Release the outer cylinder, and then pull down again. After a minute or two, milk starts to flow and collects in the outer cylinder.
7. When milk stops flowing, break the seal, pour out the milk and then repeat the procedure.
8. Take care for disinfection of equipment. Ensure strict asepsis routines are followed.

Q. 9. How does one express breast milk by hand in case breast pump is painful?

Teach the mother to do this herself. Do not express her milk for her. Touch her only to show her what to do. Be gentle.

1. Ask mother to wash her hands thoroughly.
2. Make her sit comfortably and hold the container near her breasts.
3. Ask the mother to massage the breast.

4. Ask her to put one thumb on her breast *above* the nipple and areola and her first finger on the breast *below* the nipple and areola, opposite the thumb. She supports the breast with her remaining fingers and squeezes out milk gently by using her thumb and first finger.

Q. 10. How to clean or sterilize the breast milk container?

Hospital guidelines should strictly be followed but in absence of strict hospital policy, the container can be sterilized by boiling for 10 minutes and can be cleaned each time prior to use in soap water and to be dried in air.

Section

I. Appendix

38. Steps for Equipment Selection and Writing Specification
39. List of Indian Equipment Dealers
40. List of Imported Equipment Dealers

Chapter 38

Steps for Equipment Selection and Writing Specification

Whenever we buy equipment, drawing correct specifications are the most important initial step. If specifications are not made carefully, one may end up buying third-class equipment. The specifications depend on one's individual needs. The first step is to identify the needs, then look at the detailed specifications of what equipment is available in the market and actually have a demonstration of each one of them. Only then should one draw out the final specifications. They should be as exhaustive as possible and cover the warranty and maintenance periods. We have put up some samples of the specifications of important pieces of equipment for free download here. For one's individual needs and set-up, they may be modified appropriately. For further detailed specifications look at website of National Neonatology Forum of India *www.nnfequipment.org* or National Health Systems Resource Centre *www.nhsrcindia.org*

Given below are steps for proper equipment selection and writing specifications. This is of paramount importance for proper utilization of money. Money saved will be available for buying other resource materials for the neonatal unit.

A. STEPS FOR EQUIPMENT SELECTION

The following are the essential steps for equipment selection:

1. ***Need assessment***: The reason for purchasing a particular equipment should be enunciated beforehand. In the hospitals where there is requirement of new equipment, there is a need to do the survey for justifying the need of a particular equipment. In many countries before a new facility is started a team comprising hospital administrators, hospital engineers, epidemiologist, financial experts visit the particular hospital and assess the need and thereafter they issue a certificate of need before allowing new facilities to come up.

2. ***Use coefficient***: This coefficient is applied to assess the utilization of an equipment, i.e. whether the equipment is optimally utilized or under utilized. Additional demands of the equipment may be assessed by use of the formula:

$$\text{Use coefficient} = \frac{M}{N} \times 100$$

where,
   N = Average number of hours the equipment is used per day.
   M = Maximum number of hours the equipment can be used per day.

If the use-coefficient is less than 50%, it is considered to be under utilized and hence a bad investment. *However, life-saving equipment cannot be subjected to this kind of assessment.*

3. ***Cost consciousness***: While procuring a new equipment the total cost of the equipment must be kept in mind. Cost containment procedures may be thought at every stage of the procurement procedure.

4. ***Specifications and not the brand***: There should be insistence on exact specifications of the equipment. Tenders floated with the exact specifications will set about a healthy competition which works out to advantage of user. Functional requirements should be known and based on these, technical specifications developed. One shall be able to negotiate sometimes on an unimaginative price.
5. ***CIF destination***: Normally insurance and freight are covered up to nearest port with customer clearance facility. In case of procurement of sophisticated, expensive and imported items with custom clearance may cost an additional 1–2% of the value of equipment. This should be confirmed at the time of negotiation. Most of the suppliers would be willing to absorb the additional cost and the burden of customs clearance.
6. ***Warranty with spares***: It is the age of electronics due to fierce competition in market, the manufacturers are yielding to higher and higher warranty period that too with spares. The cost of spares range from 3–5% of the value of equipment. Conventional warranty period of one year can now be bargained for 2–3 years without much additional cost. Keep in mind that every year 10–15% of the cost of equipment will be required for the spares.
7. ***Continuous supply of consumables***: Some equipment may need expensive consumables to be imported. Every effort must be able to secure as many consumables as possible to last for a minimum period of one year. Sufficient guarantee has to be ensured for continuous supply of consumables thereafter for at least 10 years.
8. ***Service contracts/after sale services***: It is important to ensure continuous and an uninterrupted functioning of the equipment. Service contract must be conceived and planned at the time of purchase. The supplier may dictate the terms and price if plan for service contract is done at the end of the warranty period. Accepted norms for the service charged are 1–2% of the cost of equipment for the first year after warranty with a 10–15% increase each year.
9. ***Training of staff***: Training of staff to handle the equipment efficiently may range from training at site to training abroad depending upon the sophistication needs of the staff and institution. This should be ensured with the supplier to provide necessary training to the staff free of charge before the order. Timely and appropriately training the staff for handling and operating the equipment is a prerequisite for effective and optimum utilization of an equipment.
10. ***Foreign exchange***: Life-saving equipment is exempted from customs duty. Foreign currency value keeps on frequently fluctuating. If it is envisaged that the lead time of the equipment to be procured is more than 3 months, it may be wiser to indulge in forward booking where bank undertake to take care of fluctuations for a specific period by charging a premium and understanding from the supplier to cover the expenses on extension of letter of credit and exchange fluctuations.
11. ***Facility for back up power supply***: As most of the vital and essential equipment are functional on electricity or chargeable battery supply, facility for back up power supply should be ensured. Some arrangements have to be made in the form of stand by generators or if possible uninterrupted power supply (UPS).
12. ***Good economics***: Equipment may be offered at a very low price but the consumables required may have to be imported later at a phenomenal cost. It may be wiser to purchase more expensive equipment which can be operated with much cheaper consumables available locally.

B. WRITING SPECIFICATIONS

Writing specifications for equipment requires in-depth knowledge about the equipment to be procured. In a private set up, where money is not the constraint, one is at liberty to buy and choose after careful scrutiny. On the other hand, in a government set up, one has to be careful while developing specifications for particular equipment for tender purposes. One need to make specifications in such a fashion, so that unreliable/untrusted dealers can be avoided. Remember often the rules will dictate to buy the lowest quotation equipment. But if the equipment does not conform to your required specifications, you are justified in rejecting a few lowest quoted equipment. Hence, it is very important that one spends time in framing the specifications of ideal equipment one is looking for. One should not compromise on the quality of equipment, even though it is higher priced. It is better not to buy junk, low priced equipment from shady dealer. Given below are a few standard specifications, we circulate for tender purposes. These can be suitably modified based on individuals need.

1. *Resuscitation bags*: The unit should consist of one self-inflating bag, one patient valve, one oxygen reservoir and 2 face masks. The bag should be made of sterilizable synthetic material like silicone. The patient valve should be well sealed, have minimum dead space and no forward or backward leaks. The oxygen reservoir should be designed to provide up to 100% oxygen. The bag should have an oxygen inlet which fits into the standard oxygen tubing both from a cylinder and central supply. Face masks should be transparent, fit the patient outlet easily and have minimum dead space. The system should withstand washing scrubbing and autoclaving procedures.

2. *Phototherapy unit*: Phototherapy unit using compact fluorescent tubes, blue and white, for use in the treatment of neonatal jaundice. It should provide irradiance of over 20–30 $\mu W/cm^2/nm$ in 420–480 nm range. It should have electronic light sensor, light focusing screen, warning LED display to indicate need for tube change. The unit should be mounted on high quality castors with locking mechanism. There should be provision for easy adjustment of height and side positioning. The unit should have electronic control for therapeutic light, no light spread and noiseless. It must run on 220–250 volts.

3. *Infant open care system with servo control*: Free standing, mobile, servo-controlled radiant warmer for neonates. It should have an overhead infrared quartz heating system with baby bed mounted on castor wheels. It should run on servo and manual modes. Provision for continuous display of temperature and heater output should be available. An observation light must be provided. There should be provision for taking X-ray. The heating unit should be provided with the facility of swiveling in order of position the X-ray hood over the baby. The baby bed should be tiltable to 5–15°.

4. *Intensive care baby incubators*: Servo-controlled intensive care newborn incubator, using double wall system. It should control baby's temperature with an accuracy of ± 0.3°C and air temperature ± 0.5°C of the corresponding set temperatures. Alarms should include those for air flow, probe failure, high incubator temperature, baby set temperature and high skin temperature (± 0.5°C). The unit should have a humidity control system (range 30–95%). The display should include baby's temperature; set temperature, heater output and humidity level. Oxygen delivery system should be in-built. The whole unit should be sleek, silent easy to move having easy access. Two skin probes should be included with each unit.

5. ***Infusion pumps***: Infusion pumps of syringe type for use with standard plastic syringes of common brands of 10, 20 and 50 ml capacity. Flow rates to range from 0.1 ml to 100 ml per hour in 0.1 ml steps with small syringes and 1 ml steps for larger syringes. There should be facilities for operation display, purging, occlusion detection and alarms (for near empty, low battery and occlusion). Battery back up for at least 90 minutes should be available. It should be possible to mount it on intravenous (IV) pole with a clamp.

6. ***Heavy duty suction machines***: Heavy duty suction machines able to run continuously 7 days a week with facility of negative suction pressure up to 760 mm Hg. The suction pressure should be adjustable with easy readable display. The collection jars should be provided with a single ball and valve mechanism, transparent, autoclavable and unbreakable. Two spare collection jars with complete tubing connections should be provided extra with each machine.

7. ***Weighing scales***: Electronic weighing scales to measure weight of neonates with an accuracy of 1 gm. Cradle for placement of the baby should be made up of preferably acrylic material. It should have easy to correct zero error. LED display should be large enough to be visible from a distance of 4–6 feet to normal eye. The unit should run on mains 220–240 volts or inbuilt battery.

8. ***Apnea monitor for newborn babies***: A small, portable, light weight unit based on the principle of picking respiratory movements. The unit should have a facility for alarm for apnea with varying duration of 10 seconds, 15 seconds and 20 seconds. The unit should be operable on rechargeable battery.

9. ***Pulse oximeter***: Pulse oximeter should detect and continuously display oxygen saturation and heart rate of the patient. It should also have the pulse display. The probes should be patient-friendly and clip type. Two extra probes should be included with each pulse oximeter. A battery back up should be available. Alarms should be available for high and low saturation and heart rate values.

10. ***End-tidal CO_2 monitor***: For measuring online end-tidal carbon dioxide on ventilated newborn. The unit should be able to measure CO_2 on very minute samples with continuous display of waveform and absolute value. The dealer must give a guarantee that periodic callibration of unit with standard CO_2 will be done by him for 3 years without extra cost.

11. ***Monitors with provision of blood pressure, oxygen saturation and ECG (with other options)***: Vital sign monitors for neonates with provision of non-invasive BP, heart rate, ECG, respiration and oxygen saturation. Optional specifications include temperature probe, invasive line (one) and centralized console facilities. It should have alarms facility for all the parameters. The display should be multicolor. Wall mounting/panel to be included. Accessories like pressure transducers, and cables to be quoted in duplicate for each monitor.

12. ***Non-invasive blood pressure monitor***: Non-invasive electronic cuff type blood pressure monitor suitable for neonates. It should display systolic, diastolic and mean blood pressure and heart rate. The monitor must have adjustable high and low alarm limits with audio and visual alarms. The instrument must have a short setting up time and an adjustable cycling time. Reusable BP cuffs of 3 sizes (newborns, infant, child) in triplicate be provided with machine. The unit must run on 220 volts main.

13. ***Ventilators***: Ventilators for neonates with time cycled, volume preset, flow generated operation, with IPPV, SIMV, pressure support and CPAP modes. Compressor to be included. Heated humidifier with servo

temperature control and reusable water chamber to be included. Display should include set values, delivered values and alarms. Graphic display of pressure-volume and flow volume loops should be optional. Alarms should be applicable to low and high values of respiratory parameters, oxygen and power failure. Circuits and tubings should be reusable (after sterilization with glutaraldehyde and autoclaving). One extra humidifier chamber (optional) to be quoted. Four additional sets of ventilator tubings should be included with each ventilator. Spares and accessories sufficient for 2 years should also be included.

14. *Transport incubator*: Transport incubator for transporting sick neonates. It should provide warmth, oxygen and suction facility. Displays should include baby's skin temperature and heater function. It should be possible to either hand carry it or move on its sturdy wheels. The battery should be easily chargeable with charge lasting 4–6 hours.

15. *Resuscitation bassinet*: Neonatal resuscitation trolley for use in the delivery room. It should have radiant warming system, examination light, stop watch, suction system and oxygen system. Efficient heating should be possible. It should provide suction of at least 130 cm of water or more.

16. *Neonatal vital signs monitors*: Monitors for critically sick neonates. It should include non-invasive BP, pulse oximeter, respiratory rate, ECG and heart rate. It should have display of all the vital parameter measured. Alarms for high and low range of all parameters should be available. The BP cuffs and pulse oximeter probes should be reusable and suitable for use in sick preterm and term neonates. All probes should be quoted in duplicate with each unit.

17. *Oxygen analyzer*: Hand-held oxygen analyzers for checking the oxygen concentration in oxygen head box and patient inlet tubing of ventilators based on teledyne/electron technology. It should run on readily available battery. It should show FiO_2 from 0.21 to 1.0 with accuracy of 1%. Alarms are optional.

18. *Glucometer*: Hand-held blood glucose monitoring devices using strips requiring only a drop of blood and giving the result in a minute or so. The equipment should be light weight and supplied with calibration standards. The glucose strips should be readily available.

19. *Bilirubinometer*: For determining the total bilirubin of neonates from capillary blood sample. It should be easy to operate and use commercially available capillary tubes. Microcentrifuge should be included as part of package.

20. *Fluxmeter*: Hand-held battery operated fluxmeter to measure and display the irradiance of phototherapy units at 425–475 nm wavelength. It should be possible to estimate the flux in microw/nm/cm^2 readily.

21. *Refractometer for urine-specific gravity*: Hand-held optical refractometer that provides bedside estimation of urine specific gravity from single drop of urine.

22. *Cold light for detection of pneumothorax*: High intensity light source for detection of pneumothorax in newborn. The unit should be light weight and portable. The tip of the probe (light source) should not get warm on being lightened. The unit should run on 220–240 volts AC.

23. *Breast pump*: Motorized breast pump unit running on mains of AC 220–240 volts with a variable suction frequency. The unit should have silicone collection cups and jars made of polycarbonate material. Four additional cups and two collection jars should be provided extra with each unit.

C. HOW TO WRITE DETAILED SPECIFICATIONS FOR A HOSPITAL TENDER?

1. *Bilirubinometer, total bilirubin, capillary based*
- Bench top point-of-care bilirubin meter—direct reading of photometer determining total bilirubin in serum/plasma
- On switch and auto-off:
   Automatic calibration setting between measurements
- *Dual wavelength measurement*: 460 nm and 550 nm
- Correcting for Hb at 550 nm
- *Sample size*: 1 capillary tube with serum/plasma
- Main light source, 5 W tungsten lamp
- Measuring range: 0 to 700 µmol/or 0 to 40 mg/100 ml
   Accuracy equivalent to laboratory spectrophotometer (approx ± 5%)
- Read-out switchable between mg/100 ml of µmol/L
- Fast analysis time <5 sec
- Large LED display readable in low light working situations and display cover durable plastic
- With integrated printer
- *Power requirements*: 220 V/50 Hz (with adapter)
- *Power consumption*: 350 W
- Device is produced by ISO 9001 certified manufacturer (certificate to be submitted, further details *see* "technical provisions")
- Device is safety certified according CE 93/42, FDA 510 k or equivalent (certificate to be submitted, further details *see* "technical provisions").

Supplied with
- 2 × reference solution packages
   – 1 × box of microcapilary tubes, inner diameter 1 mm, length 7 mm, heparinzed
   – 1 × pack of sealing compound for microcapillary tubes
- 1 × spare lamp
- 1 × dust cover
   1 × spare set of fuses
- User manual with troubleshooting guidance in English
- Technical manual with maintenance and first-line technical intervention instructions in English
- List of priced accessories
- List of priced spare parts
- List with name and address of technical service providers in India
- Training and installation at end-user site
- Proposal for full service AMC, years 1 to 5, covering (i) 2 preventive maintenances per year, (ii) on-call technical interventions, spare parts and travel.

2. *Pulse oximeter, bedside, neonatal*: Compact portable bedside pulse oximeter with LCD display.
   Continuous monitoring of SpO_2 (arterial blood oxygen saturation), pulse rate and signal strength.

Measuring range

SpO_2: 30 to 100%, min graduation 1%

Pulse rate: 20 to 250 bpm, minimal graduation 1 bpm

Accuracy SpO_2: 50 to 69% (± 3%), 70 to 100% (± 2%)

Display shows SpO_2(%), HR (bpm) and signal strength bar

- Large display readable from distance, display cover durable plastic:
   – User preset of high/low alarms on SpO_2 and pulse rate monitoring
   – Audiovisual alarm for SpO_2 and pulse rate
   – Silencing feature for audio alarm
   – Display reports system errors, probe failure and built-in battery status
   – Automatic switch from mains to batteries in case of power failure

Power requirements: 220 V/50 Hz and internal re-chargeable battery (autonomy approx 6 hrs, automatic recharge)
Power consumption: 50 W
- Device is produced by ISO 9001 certified manufacturer (certificate to be submitted, further details *see* "technical provisions")
- Device is safety certified according CE 93/42, FDA 510 k or equivalent (certificate to be submitted, further details *see* "technical provisions").

Supplied with
2 × reusable SpO$_2$ sensors neonate, clip-on type (including connection cable)
- 10 × reusable SpO$_2$ sensors neonate, wrap around type (including connection cable):
  - 1 × spare rechargeable battery
  - 1 × spare set of fuses
- User manual with troubleshooting guidance in English
- Technical manual with maintenance and first line technical intervention instructions in English
- List of priced accessories
- List of priced spare parts
- List with name and address of technical service providers in India
- Training and installation at end-user site
- Proposal for full service AMC, years 1 to 5, covering (i) 2 preventive maintenances per year, (ii) on-call technical interventions, spare parts and travel.

3. *Scale, baby, electronic, weighing up to 10 kg*
- Electronic scale for weighing babies
- Measuring range 0 to approx 10 kg
- *Minimum graduation*: 5 g
- With tare function
- On switch and auto-off
- Autocalibration with each scwitch-on
- Large LED display readable in low light working situations, display cover durable plastic
- Display in kg and lbs, easy switch between kg and lbs

- Reading time max 5 seconds
- Zero weighing adjustment
- Freeze reading feature
- Smooth surface/finishing allows for easy cleaning/disinfection
- All vital parts made of rust proof materials
- Horizontal levelling with height adjustable feet
- Splash proof and shock resistant light-weight body
- *Power requirements*: 220 V/50 Hz
- *Power consumption*: 150 W
- Device is produced by ISO 9001 certified manufacturer (certificate to be submitted, further details *see* "technical provisions")
- Device is safety certified according CE 93/42, FDA 510 k or equivalent (certificate to be submitted, further details *see* "technical provisions").

Supplied with
- 1 × spare set of fuses
- User manual with troubleshooting guidance in English
- Technical manual with maintenance and first line technical intervention instructions in English
- List of priced accessories
- List of priced spare parts
- List with name and address of technical service providers in India
- Training and installation at end-user site

4. *Syringe pump, 10, 20, 50 ml, electrical, 220 V*:
- Digital and self-regulating volume controlled portable syringe pump
- Can be mounted on standard bed/wall rail or mobile pole/stand (supplied with fixation)
- Suitable for all intravenous and intra-arterial infusions
- Continuous volumetric delivery with syringes 10, 20 and 50 ml
- Open system, suitable for different brands of syringes
- Programmable, user entry: Infusion volume and time or flow rate

- *Rate, adjustable*: 1 to 999 ml/hr, steps of 1 ml/hr
- *Accuracy*: Ca 1% of total volume delivered
- With occlusion detection and alarm
- Display reports systems errors, end of infusion and built-in battery status
- Audiovisual alarm with silencing feature for audio alarm
- Automatic switch from mains to batteries in case of power failure
- *Power requirements*: 220 V/50 Hz or internal re-chargeable battery (autonomy approx 6 hrs, automatic recharge)
- *Power consumption*: 50 W
- Device is produced by ISO 9001 certified manufacturer (certificate to be submitted, further details *see* "technical provisions")
- Device is safety certified according CE 93/42, FDA 510 k or equivalent (certificate to be submitted, further details *see* "technical provisions").

Supplied with
- 1 × spare battery
  1 × spare set of fuses
- User manual with troubleshooting guidance in English
- Technical manual with maintenance and first line technical intervention instructions in English
- List of priced accessories
- List of priced spare parts
- List with name and address of technical service providers in India
- Training and installation at end-user site
- Proposal for full service AMC, years 1 to 5, covering (i) 2 preventive maintenances per year, (ii) on-call technical interventions, spare parts and travel.

5. *Phototherapy unit, single head, high intensity*
- Technical specifications
  – Heavy sturdy mobile stand phototherapy unit
  – Antistatic castors, 2 with breaks
- *Single head, surface size, approx*: 0.50 × 0.75 m
- *Head height adjustable, approx*: 1.40 to 1.75 m
  – Blue light, 4 compact fluorescence tubes (CFL), 20 W
  – White light, 2 compact fluorescence tubes (CFL), 20 W
  – Tubes are protected by grill
- *Irradiance at skin level, up to*: 30–40 µW/cm^2/nm
- *Wavelength*: 420 to 500 nm, with highest intensity at 470 nm
  – Integrated cumulative hour timer
- *Power requirement*: 220 V/50 Hz
- *Power consumption*: 250 W
- Device is produced by ISO 9001 certified manufacturer (certificate to be submitted, further details *see* "technical provisions")
- Device is safety certified according CE 93/42, FDA 510 k or equivalent (certificate to be submitted, further details *see* "technical provisions").

Supplied with
- 10 × spare blue CFL tubes
- 4 × spare white CFL tube
  – 1 × spare set of fuses
- User manual with troubleshooting guidance in English
- Technical manual with maintenance and first-line technical intervention instructions in English
- List of priced accessories
- List of priced spare parts
- List with name and address of technical service providers in India
- Training and installation at end-user site
- Proposal for full service AMC, years 1 to 5, covering (i) 2 preventive maintenances per year, (ii) on-call technical interventions, spare parts and travel.

6. *Open care system on trolley with drawers, with radiant warmer, O_2-provision*
- Mobile newborn resuscitation table with fixed-height radiant warmer

- Antistatic castors, 2 with breaks
- Table surface with mattress with infant head/shoulder support
- *Mattress-padding*: Foam density approximately 21–25 kg/m^3
- Mattress cover: Removable with zipper, waterproof, washable, resistant to cleaning with chlorine based solution and flame retardant
- Side boards transparent acryl, drop down and lockable
- Under table 2 storage drawers
- Side rails allow for mounting of accessories
- Hood suspended above the table integrates heating element and overhead light
- *Overhead light*: 2 × 50 W halogen spot, with dimming function
- Integrated support for two 10 L oxygen bottles
- Control unit has flow meter and displays pressure
- Heating element: Emitter with parabolic reflector and protected by metal grid
- Control unit allows air and skin temperature preset (LED indicator) and drives radiant heater output (servo and manual)
- *Integrated timer*: 1 to 59 min, with count-up and count-down feature
- *Temperature range, skin*: 34 to 38°C (user pre-settable)
- Monitoring of skin temperature by means of sensor, range: 30 to 42°C
- *Heater output*: 0 to 100% in increments of 5%
- *Control unit*: Audiovisual alarms according to timer and temperature presets avoiding overheating
- Display reports systems errors, sensor failure
- *Power requirement*: 220 V/50 Hz
- Power consumption: 800 W
- Device is produced by ISO 9001 certified manufacturer (certificate to be submitted, further details *see* "technical provisions")
- Device is safety certified according CE 93/42, FDA 510 k or equivalent (certificate to be submitted, further details *see* "technical provisions").

Supplied with
- 1 × mattress
- 2 × skin temperature probe (including connection cable)
- 2 × spare skin temperature probe (including connection cable)
- 1 × spare heating element
- 2 × empty 10 L oxygen cylinders
  – 1 × spare set of fuses
- User manual with troubleshooting guidance, in English
- Technical manual with maintenance and first line technical intervention instructions in English
- List of priced accessories
- List of priced spare parts
- List with name and address of technical service providers in India
- Training and installation at end-user site
- Proposal for full service AMC, years 1 to 5, covering (i) 2 preventive maintenances per year, (ii) on-call technical interventions, spare parts and travel.

For more detailed specifications visit www.nhsrcindia.org

Chapter 39

List of Indian Equipment Dealers

S.No.	Addresses
1.	Agilent (Hewlett-Packard India Ltd.) Paharpur Business Centre 21, Nehru Place New Delhi–110019
2.	Advance Mediplus P. Ltd. M-59, Gali No. 3, Shastri Nagar New Delhi–110 052
3.	Advanced Micronic Devices Ltd. 907, Kailash Building 26 Kasturba Gandhi Marg New Delhi–110001
4.	Allied Medical Limited 76-77 Udyog Vihar, Phase IV Gurgaon–122015
5.	A&R Healthcare Pvt. Ltd. DH-204, Abhimanyu Apartments Vasundhara Enclave Delhi
6.	Avery India Ltd. 8, NM Marg Ballard Estate Mumbai–400038
7.	AVL Biomedical Pvt. Ltd. C/9, Vasant Kunj New Delhi–110070
8.	Baxter (India) Pvt. Ltd. 5th Floor, Enkay Towers Udyog Vihar, Phase V Gurgaon–122106
9.	B. Braun Medical (India) Pvt. Ltd. Unit No. 1, 5th Floor, East Quadrant The IL and FS Financial Centre Bandra Kurla Complex, Bandra (East) Mumbai–400051
10.	Bernard Surgical and Scientific 26, West Patel Nagar New Delhi–110008
11.	Bergen Healthcare Pvt. Ltd. 305-306, Magnum House-I Karampura Commercial Complex New Delhi–110015
12.	Bird Meditech D/143, Jai Bonanza Indl. Est. Ashok Chakravarty Road, Kandivli (E), Mumbai–400 101
13.	BOC India Ltd. 48/1, Diamond Harbour Road Calcutta–700027
14.	BPL India Ltd. BPL Centre, 32, Church Street Bangalore–560001
15.	Cardio Products Corporation 504, Vishal Bhawan 95, Nehru Place New Delhi–110019
16.	Cardiotrace Electronics 411 Somdatt Chambers-II 9 Bhikaji Cama Place New Delhi–110066
17.	Care Fusion Care International Medical Division N 19 Green Park Extn. New Delhi–110016
18.	CL Micromed Pvt Ltd. C-123 Lajapat Nagar New Delhi–110024

List of Indian Equipment Dealers

S.No.	Addresses
19.	C-MEC (India) Post Box 486, Thycaud P.O. Trivandrum–695014
20.	Consolidated Products Corp. Pvt. Ltd. F-1/8, Okhla Industrial Area, Phase-I New Delhi–110020
21.	Core Healthcare Limited Core Towers, Near Parimal Crossing Ellisbridge Ahmedabad–380006
22.	Criticare Systems India 507-508, Vishal Tower District Centre Janakpuri New Delhi–110 058
23.	Delhi Hospital Supply Private Ltd. 101-103, Pal Mohan Sadan 26/32, East Patel Nagar, Rajendra Place New Delhi–110008
24.	Delhi Surgical and Dressings Pvt. Ltd. No. 6, Moti Cinema Compound PB No. 1339, Chandni Chowk Delhi–110006
25.	Denis Meditek 706, Ansal Chamber-II 6, Bhikaji Cama Place New Delhi–110066
26.	Delta Medical Appliances 108, Damji Shamji, Industrial Complex Mahakali Caves Road, Andheri (East) Mumbai–400029
27.	Edwards Lifesciences (I) Pvt. Ltd. 7/27, Kirti Nagar, Okhla Industrial Area New Delhi–110015
28.	Elder Health Care 11-G, Gopala Towers 25, Rajendra Place New Delhi–110008
29.	Draeger Medical India Pvt. Ltd. B-25, Lala Lajpat Rai Marg Lajpat Nagar-II New Delhi–110 024
30.	EMCO Meditek Private Limited 106, Industrial Area, Sion Mumbai–400022
31.	Fanem Medical Devices India Private Ltd., Manufacturer Address: 279, 4th Main Road, Ganapathy Nagar, Peenya, Bengaluru Karnataka–560058
32.	Fisher & Paykel Healthcare India Pvt. Ltd. India Branch Office, 339/1 HIG A Sector, 2nd Stage Extension, Yelahanka, New Town Bangalore–560 064
33.	Fresenius Kabi India Ltd. Heritage House 6-E, Ramabai Ambedkar Road Pune–411 001
34.	Global Medical Systems B1/34, Ground Floor, Model Town-II Delhi–110009
35.	Grabner International 7/1, West Patel Nagar New Delhi–110008
36.	Helix Corporation # 878, 17th Cross, 9th B Main Road, ISRO Layout Bangalore–560078 (India)
37.	Hera Global Marketing and Research B-134, Taimoor Nagar, Opp C Block New Friends Colony New Delhi–110065
38.	Hicks Thermometers C-26, Industrial Estate Aligarh–202001
39.	Hicon Eminence 28, Lehna Singh Market, Malka Ganj Delhi–110 007
40.	Hudson RCI CNC Medical Devices 262 Adi Mansion, Dr Cawasji Hemusji Street, Mumbai–400002

S.No.	Addresses
41.	Indchem Med. System Ltd. 47, Developed Plots for Electronics Ind., Perungudi Chennai–600096
42.	Indian Surgical Equipment Co. Pvt. Ltd., C-6, Hauz Khas New Delhi–110016
43.	Innovative Intex Pvt. Ltd. 1004, New Delhi House 27 Barakhamba Road New Delhi–110001
44.	Instruments and Machine Inc. 3, Arugmuga Naiken Street, Main Road Chennai–600002
45.	International Surgico Industries 3071/8, Partap Street Behind Golcha Cinema, Darya Ganj New Delhi–110002
46.	Instromedix (India) Pvt. Ltd. Pragati Chambers, Ranjit Nagar Commercial Complex New Delhi–110008
47.	J Mitra and Company A-180, Okhla Industrial Area Phase-I, New Delhi–110020
48.	Kardio Control Co. Inter Kardio Pvt. Ltd. A-23, Wazirpur Indl. Area Delhi–110052
49.	Ferdal Global Health India Fl2, Abhishek Plot 178 Sr No 96/97 Rt Bhusari Colony Kothrud, Pune–411038
50.	Larsen and Toubro Limited Medical Equipment and Systems Gate No. 7, Annexe Building 2nd Floor, PB No. 8901 Saki Vihar Road, Powai Mumbai–400 072
51.	Lectromedik Ltd. 35/52, Industrial Subrub, Yeshwantpur Bangalore–560022

S.No.	Addresses
52.	Masimo Medical Technolgies India Pvt. Ltd. 70/2 Millers Road Miller's Boulevard, 2 Floor Bangalore–560052
53.	Medex India (P) Ltd. DD-36, Kalkaji New Delhi–110019
54.	Medi Equip India Pvt. Ltd. Flat No. 104, 9/2 East Patel Nagar New Delhi–110008
55.	Medical Co-ordinators Pvt. Ltd. No. 8, Opp. B-2, Nizamuddin (West) New Delhi–110013
56.	Medical Systems and Services (North Contact for Meditrin) H-251, Ashok Vihar, Phase-I Delhi–110 052
57.	Mediland, 41-42, Liberty Cinema, Marine Lines Mumbai – 400002
58.	Mediserve H-251, Ashok Vihar, Phase-1 New Delhi–110052
59.	Medisphere Marketing Ltd. 201, 28-29, Link Road, Feroze Gandhi Marg Lajpat Nagar, Part-III New Delhi–110024
60.	Medisys 118-119, Shriram House Ashram Crossing New Delhi–110014
61.	Meditrin Madhuvan 98, Patharewadi Opp. Bageecha Restaurant Malvani Church Marve Road Malad (W) Mumbai–400 095

List of Indian Equipment Dealers

S.No.	Addresses
62.	Miles India Ltd. 589, Sayajipura, Ajwa Road Baroda–390006 Mindray Medical India Pvt Ltd Unit no. 401/402, NDM-1 Netaji Subhash Place Wazir District Centre New Delhi–110034
63.	Mira Cradle 610–A, Udyog Vihar Phase-I, Gurgaon, Haryana–122016
64.	Moolaa Technologies Pvt. Ltd. No. 451, 9th Cross JP Nagar, II Phase Bangalore–560078
65.	Nautek Service Pvt. Ltd. 402, Gagan Deep 12, Rajendra Place New Delhi–110008
66.	Nice Neotech Medical Systems (P) Ltd. No. 1. (159) Pallavan Street Alwarthirunagar Chennai–600 087
67.	Nihon Kohden India Pvt. Ltd. 308, Tower-A, Spazedge Sector–47, Sohna Road, Malibu Town Gurgaon, Haryana–122002
68.	Nu Tek Instruments 6, BR Complex, C-32, Patparganj Mayur Vihar, Delhi–110091
69.	Osaw Industrial Products Pvt. Ltd. PO No. 42, OSAW Complex, Jagadhri Road, Ambala Cantt–133 001
70.	Pharmamen Enterprises Pvt. Ltd. E-107 Lajpat Nagar –I New Delhi–110024
71.	Philips Medical Sys. India Ltd. A-16, Ground Floor Mohan Co-operative Industrial Estate Ltd. Mathura Road New Delhi–1100044

S.No.	Addresses
72.	Phoenix Medical Sys. Pvt. Ltd. No. 32/4, Jawaharlal Nehru Salai Ekkattuthangal, PB No. 3205 Chennai–600 097
73.	Pulmoncare Consultants E-202, LGF, Greater Kailash, Part-II New Delhi–110048
74.	Regisar Sons 42, LSC A-Block, Naraina, Ring Road New Delhi–110028
75.	Respirotech Med Solutions Pvt. Ltd. Room No. 201, First Floor Amar Plaza, 45A Hasanpur IP Extension New Delhi–110 092
76.	Rohanika Electronics and Med. Sys. A-207, Sarita Vihar New Delhi–110 076
77.	Rohit Surgical Pvt. Ltd. 9/27, East Patel Nagar Opp. Hotel Siddhartha, Rajendra Place New Delhi–110008
78.	RPPL TM 5012-14, Guru Nanak Marg Ambala–133 001, Haryana, India
79.	Rustagi Surgical Pvt. Ltd. 1/2571, Ram Nagar Loni Road, Shahdara Delhi–110032
80.	Sandeep Instruments and Chemicals, 3229, Ranjeet Nagar New Delhi–110 008
81.	SBP Medicare Systems Pvt. Ltd. J-3/16 (Opp Happy School) Daryaganj New Delhi–110 002
82.	Schiller Healthcare India Ltd. Advance House, Makwana Road Andheri (E), Mumbai–400059
83.	Scorpia India Medicare Pvt. Ltd. 3/N/6/1 Chaudhary Moad Ambedkar Road, Ghaziabad, UP

Neonatal Equipment

S.No.	Addresses
84.	Shreeyash Electo Medicals Sr. No. 49, Plot No. 22 Shri Sai Industrial Estate Gujarwadi Phata Katraj Pune–411 046
85.	Siemens Ltd. Medical Engineering Division 4-A, Ring Road, IP Estate New Delhi–110002
86.	Sono Site Inc Vatika Business Centre First India Place MG Road Gurgaon–122002
87.	Surgifield 31, Beadon Street Calcutta–700006
88.	Super Medicare Agencies RZ-130, Block-A, Sitapuri, Part-I Palam Road New Delhi–110045
89.	Systems Biomedical Pvt. Ltd. 103, Pankaj Chambers, Preet Vihar Community Centre Delhi–110092
90.	Systronics, 89-92, Industrial Area, Naroda Ahmedabad–382 330, (Gujarat)
91.	S S Medicaid E-270, Shastri Nagar Delhi–110052
92.	SS Technomed Private Limited 20 F-Block, Local Shopping Centre Ashok Vihar, Phase-I Delhi–110 052
93.	Terumo Corporation, Alexander Square 4th Floor, 34 and 35, Sardar Patel Road Guindy, Chennai–600 032
94.	Trans Health Care India Pvt. Ltd. AP 740, 15th Sector, 95th Street KK Nagar, Chennai–600 078
95.	Trivitron Medical Systems Pvt. Ltd. B-1/A-12 Mohan Cooperative Ind Estate, Mathura Road New Delhi–110 044

S.No.	Addresses
96.	Tekno Equips 1/4, Lazzar Road Cross Bangalore–560005
97.	Towa Optics (Ind.) Daryaganj Delhi–100092
98.	Twenty First Century Medicare Ltd. TCML House, 23 Pushp Vihar Community Centre New Delhi–110062
99.	Ultra Clean Systems B-41/C (Behind Govt. School) Sarai Kale Khan, Nizamuddin New Delhi–110013
100.	Usha Draeger Ltd. B-II/94, MCIE Sher Shah Suri Marg Near Badarpur Police Station New Delhi–110044
101.	Unissi (India) Pvt. Ltd. F-3 (B), NDSE, Part-II New Delhi–110 049
102.	Ved Mediserve Pvt. Ltd. 101, 201, Exp Bldg, H-Block, Ashok Vihar, Phase I Delhi–52
103.	Ved Med Software and Trading Pvt. Ltd 114, Sant Nagar, East of Kailash New Delhi–110055
104.	VIASYS C/o Life Care Medical Systems 9-A Vanrai, Vishwakarma Nagar Nahur Road, Mulund (West) Mumbai–400080
105.	Vishal Surgical Equipment Company Ashoka Chamber, B-5, Rajendra Park Pusa Road, New Delhi–110060
106.	Wipro GE Healthcare 249 1st Floor, Okhla Indistrial Area Phase I New Delhi–110020
107.	Zeal Med Pvt. Ltd. 4/19-A, Piramal Nagar Indl. Est. SV Road, Goregaon (W) Mumbai–400062

Chapter 40

List of Imported Equipment Dealers

S.No.	Addresses
1.	Artema Medical AB, Rissneleden 136 S-174 57 Sundbyberg, Sweden www.artema.dk Indian Dealer: Trivitron
2.	Atom Medical Corp 3-18-15, Hongo, Bunkyo-ku, Tokyo, Japan www.atomed.co.jp Indian Dealer: Vishal Surgical Equip. Co. Pvt. Ltd.
3.	B. Braun USA www.bbraunusa.com Indian Dealer: B. Braun Medical (I) Pvt. Ltd.
4.	Baxter Access Systems Route 120 & Wilson Road Round Lake, IL 60073, USA www.baxter.com Indian Dealer: Baxter (India) Pvt. Ltd.
5.	BCI Inc. N7 W22025, Johnson Drive Waukesha, Wisconsin 53186, USA www.smiths-bci.com Indian Dealer: Rohanika Electronic and Medical Systems
6.	Care Fusion 1605B, Sino Plaza, Hongkong www.carefusion.com
7.	Criticare Systems Inc. 20925 Crossroads Cir., Suite 100 Waukesha, WI 53186-4054, USA www.csiusa.com Indian Dealer: Criticare Systems India
8.	Datex-Ohmeda Instrumentarium Corp., Teollisuuskatu 29 PB No. 900, Helsinki, FIN 00031 Finland www.ohmedamedical.com Indian Dealer: Datex Ohmeda Pvt. Ltd.
9.	Devil Biss Healthcare 100 Devilbiss Drive, Somerset, PA 15501 www.DevilBissHealthcare.com
10.	Draeger Medical AG & Co. Kga A, Moislinger Allee 53-55 23542, Lubeck, Germany www.draeger.com Indian Dealer: Draeger Medical
11.	Edwards www.edwards.com
12.	EME Ltd. 60 Gladstone Place, Brighton Sussex BN2 3QD, ENGLAND www.eme-med.co.uk Indian Dealer: Criticare Systems Inc.
13.	Fanhem Sao Paulo-Brasil Av.Gal.Ataliba Leonel 1790CEP 02033-020-Sao Paulo–SP Brasil www.fanem.com.br

S.No.	Addresses
14.	Fisher & Paykel Healthcare 15 Maurice Paykel Place East Tamaki, PO Box 14348 Panmure, Auckland 6, New Zealand www.fphcare.com Indian Dealer: Fisher and Paykel Healthcare
15.	Fresenius Kabi Hong Kong Ltd. Room 5101-23, 51/F Sun Hung Kai Centre, 30 Harbour Road, Wanchai Hong Kong www.www.fresenius-kabi.com Indian Dealer: Fresenius Kabi India Ltd.
16.	Galemed Corpn. 10F, No. 109, Sec 6 Min Chiuan E Road, Taipei City, Taiwan www.galemed.com Indian Dealer: Rustagi Surgicals Pvt. Ltd.
17.	Ginevri Technologie Biomediche, Ginevri srl via Cancelliera, 25/D-00040 Cecchina (Roma) Italy www.ginevri.com Indian Dealer: Global Medical Systems
18.	Hill-Rom 1069 State Route 46 East Batesville, IN 47006, USA www.hill-rom.com Indian Dealer: Hill-Rom ME
19.	International Biomedical Inc. Airborne Life Support Systems Div 8508 Cross Park Dr, Austin, TX 78754-4532, USA www.int-bio.com Indian Dealer: Medisphere marketing Ltd.
20.	Laerdal Medical Corp. 167 Myers Corners Road Wappingers Falls, New York 12590, USA www.laerdal.com Indian Dealers: laerdal Global Health India, Fl2, Abhishek Plt 178 Sr No 96/97 Rt Bhusari Colony Kothrud, Pune-411038 Delhi Surgical and Dressings Pvt. Ltd.
21.	Masimo Corp. 2852, Kelvin Ave. Irvine, CA 9614, USA www.masimo.com Indian Dealer: Masimo India Pvt Ltd
22.	Medela AG Medical Technology Lattichstrasse 4b 6341 Baar, Switzerland Indian Dealer: Rohit Surgical Pvt. Ltd.
23.	Newport Medical Instruments Inc. PB No. 2600, Newport Beach, CA 92658, USA www.newportnmi.com Indian Dealer: Pulmocare Consultants
24.	Nihon Kohden Corp. 1-31-4 Nishiochiai Shinjuku-ku Tokyo 161–8560 www.nihonkohdon.co.jp Indian Dealer: Nihon Kohden India Pvt. Ltd.
25.	Nonin Medical Inc. 2605, Fernbrook Lane N Plymouth, MN 55447, USA www.nonin.com Indian Dealer: 1. Innovative Intex Pvt. Ltd. 2. Respicure Medsys Pvt. Ltd.
26.	Novametrix Medical Systems Inc. 5 Technology Drive, Wallingford CT 06492–1926, USA www.novametrix.com Indian Dealer: Rustagi Surgicals Pvt. Ltd.
27.	Pacetech Medical Monitors 510, Garden Avenue No. Clearwater, FL 33755, USA www.pacetech-med.com Indian Dealer: Medex India (P) Ltd
28.	Palco Labs 8030 Sequel Avenue, Suite 104 Santa Cruz, CA 95062-2032, USA www.palcolabs.com Indian Dealer: Global Medical Systems

List of Imported Equipment Dealers

S.No.	Addresses
29.	Radiometer A/S Akandevej 21, DK-2700 Bronshoj, Denmark www.radiometer.com Indian Dealer: 1. SBP Medicare Systems Pvt. Ltd. 2. Trivitron Diagnostics Ltd (East and South) 3. Analytical Automation (I) Pvt. Ltd. (West)
30.	SECA 40 Barn Street B5 5QB Birmingham United Kingdom www.seca.com Indian Dealer: Scorpia India
31.	Sechrist Industries Inc. 4225 E, La Palma Ave. Anaheim, CA 92807, USA www.sechristind.com Indian Dealer: Pulmocare Consultants
32.	Shenzhen Mindray Biomedical Electronics Co. Ltd., Mindray Building, Keji 12th Road South Nanshan, Shenzhen 518057 People's Republic of China www.mindray.com.cn Indian Dealer: Rustagi Surgicals Pvt. Ltd.
33.	SLE Twin Bridges Business Park 232 Selsdon Road, South Croydon CR26PL, UK www.sle.co.uk Indian Dealer: 1. SBP Medical Systems Pvt. Ltd. 2. Phoenix Medical Systems Pvt Ltd
34.	Smart-C International Ltd. 5F-3, No. 167, Sec. 1, XinTai 5th Road Xizhi City, Taipei Hsien 221 Taiwan ROC smartci@so-net.net.tw Indian Dealer: Medi-Aid Systems
35.	Spacelabs Medical, 5150, 220th Av SE, PB No. 7018 Issaquah, WA 98027-7018, USA www.spacelabs.com Indian Dealer: Datex Ohmeda Pvt. Ltd.
36.	Stephan GmbH, Medizintechnik Kirchstrasse 19, Gadenbach Germany D-56412 www.stephan-gmbh.com Indian Dealer: Schiller India
37.	Terumo Medical Corporation 44-1,2-Chome, Hatagaya Shibuya-ku, Tokyo 151-0072, Japan www.terumo.co.jp Indian Dealer: Terumo Corporation
38.	VIASYS Healthcare, 1100 Bird Center Drive Palm Springs CA 92262, USA www.viasyshealthcare.com Indian Dealer: 1. Life Care Medical Systems 2. Rohanika Electronics & Med Systems
39.	Wipro GE Medical Systems USA www.gemedicalsystems.com Indian Dealer: Wipro GE Medical Systems Pvt. Ltd.

Index

Abnormal trace EEG 88
Advanced neonatal ventilator 200
Ambulatory pumps 120
Amplitude modulation 82
Apnea monitors 43
Argyle prongs 188
Assist/control ventilation 202
Auscultatory method 50
Automatic temperature compensation 153
Axial scans 83

Bellows 220
BERA phone
    automated auditory brainstem evoked
      response audiometry 94
Bili-blanket 116
Bililrubin 154
Bili-timer 114
Binasal
    flow oxygen cannula 192
    prongs 193
Blood pressure monitors 48
Blood pressure monitoring technique
    auscultatory 50
    Doppler 50
    finger plethysmography 51
    flush 50
    invasive 51
    oscillometry 51
    palpatory 50
Brand of equipment 5
Breast pump
    relax phase 224
    release phase 225
    suck phase 225
Brightness modulation 82
Bronze baby syndrome 114
Bubble CPAP 187
Buying equipment 6

Capnogram 76
Capnograph (end-tidal carbon dioxide monitors) 75
Capnography 77
Cardiogenic oscillations 79

Centrifuge 159
Cerebral function monitor 87
    clinical applications 90
Chemiluminescence analyzers 210
Cling wrap 12
Clinical applications of CFM 90
Compact fluorescent tubes 110
Configurational isomerization 107
Convex transducer 84
Cooling equipment
    for therapeutic hypothermia 122
Co-oximetry 35
Coronal section 83
CPAP 185
    Bubble CPAP 187
    Delivery system 185
    Hudson 189
    Prongs 188

Diastolic blood pressure (DBP) 48
Diazo method 155
Digital electronic thermometers 59
Dioctylphthalate in HEPA filter 169
Distortion product 100
Doppler ultrasound techniques 50
Draeger ventilator 200
Drip rate pumps 118
Dual wavelengths 55

End-tidal CO_2 75
Electronic thermometers 59
Equipment
    availability 6
    brand 5
    for biochemical and laboratory measure 143
    imported 4
    maintenance and service 6
    monitoring therapy 130
    selection and writing specification 231
    therapy/treatment purposes 104
    thermal control 8

Face masks 186
Fiberoptic pads 110
Filtration 164

Finger plethysmography 51
Flow
    driver 189
    generator 189
    splitter 181
Fluorescent lamp devices 109
Fluxmeter (irradiance meter
  for phototherapy unit) 136
Foot suction 220

Glucose estimation
    instruments 144
Glucose oxidase method 145
Gravity controlled devices 117

Halogen spotlights 109
Heated humidified high flow
  nasal cannula (HHHFNC) therapy 195
Hexokinase method 145
High efficiency particulate air (HEPA) filter 164
Humidification 19
Humidifier 189
Hyperventilation 80

Illuminator 115
Imported equipment dealers 245
Incubators 17
    transport 25
Indian equipment dealers 240
Indications for phototherapy 108
Indigenous equipment 3
Inductive plethysmography 45
Infrared
    detector 75
    sensing skin thermometers 59, 63
    thermometry 59
Infusion pumps
    ambulatory 120
    drip rate 118
    gravity controlled 117
    multi-channel 120
    positive displacement 118
    smart 120
    syringe 118
    volumetric 118
Inhaled nitric oxide delivery systems 205
Intensive phototherapy 115
Intermittent temperature monitoring 59

Laboratory microcentrifuge 159
Laminar airflow
    rooms/hoods 166
    system 164

LCD skin thermometer 63
Life-saving equipment 170
Light emitting diodes (LED) 110
Linear array transducers 84

Magnetometer 45
Masimo
    rainbow SET technology 42
Mass spectrometer 75
Mean blood pressure (MBP) 48
Medical equipment accreditation 4
Mercury in glass thermometer 59
Methemalbumin 154
Methemoglobinemia 210
Microcentrifuge 159, 160
Movement sensors 44
Multi-channel pumps 120

Nasal prongs 186
Nasopharyngeal
    prongs 186, 188
    tube 186
Neonatal equipment 3
    procurement 3
    use and maintenance 3
    ventilators 198
NIBP monitor 50

Oscillometric technique 51
Otoacoustic emission (OAE)
    distortion product 100
    machine 97, 99
    transient evoked 100
Oxygen
    analyzer 132, 133
    concentrator 179
    delivery 213
    inlet 173
    reservoir 174
Oxyhemoglobin 154

Pacifier thermometer 63
Passive apnea monitor 45
Patient triggered ventilation 202
Photo-oxidation 107
Phototherapy
    indications 108
    light sources 108
    mechanism 107
    units 106, 107
Physics of pulse oximetry 33

Physiology of thermoregulation in newborn 17
Plasticizer 169
Practical working of pulse oximeter 34
Pressure
    gauge 174
    release valve 175
    support ventilation (PSV) 203
Principle of spectrophotometry 154
Prongs CPAP
    Argyle 188
    Fisher 188
    Hudson 188
    Nasopharyngeal 188
    Paykel 188
PTV 202
Pulse oximeter
    complication 37
    physics 33
    practical working 34
    pulse oximetry vs co-oximetry 35
    usefulness 35
Pulse oximetry complications 37

Radiant warmers 11
    advantages and disadvantages 16
Re-breathing 80
Refractometer 150
Relax phase 224
Release phase 225
Respiratory inductive plethysmography 45
Resuscitation
    masks 175
    therapy 213
    trolley 213

Sagittal scan 83
Sector transducer 84
Self-inflating bag 173
    manual resuscitator 172
Serum bilirubin estimation 154
SET technology 42
Skin probes application 12, 15
Smart infusion pumps 120
Spectral reflectance meters 55
Spectrometric bilirubin analyzer 154
Spectrophotometry
    estimation of bilirubin 154
Structural isomerization 107
Suck phase 225

Suction
    machine 220
    unit 213
Synchronized intermittent mandatory ventilation 202
Syringe pumps 118
Systolic blood pressure (SBP) 48

Tecotherm Neo 125
Temperature control modes 19
The Blanketrol III 124
Therapeutic hypothermia 122
Thermistor probe 22, 61
Thermocouple probes 22, 61
Thermocycling 169
Thermometers 59
    infrared sensing 63
    LCD skin 63
    pacifier 63
Thermoregulation 17
Thoracic impedance monitors 44
T-piece resuscitator 216
Transcutaneous
    bilirubinometer 55
    blood gas monitors 68
    carbon dioxide ($TcPCO_2$) monitors 71
Transilluminator 139
Transport incubators 25
Turbidity 154
Types of
    centrifuge 159
    refractometer 151
    transport incubators 25
    weighing scales 65

Ultrasound machine 82
    A mode 82
    B mode 82
    M mode 82
Use of available equipment 6
    warmer 12
Usefulness of pulse oximetry 35

Valve assembly 173
Variable/dual flow CPAP 189
Ventilators 198
Volumetric pumps 118

Weighing scales 64
    electronic (preferable) 65
    spring balance 65
White tubelights 113
Whole body cooling 124

Note

Note

Note

Note

Note

Note

Note